SECOND EDITION

PREHISTORIC LIFE

AN EXAMINATION OF THE HISTORY OF LIFE AND EVOLUTION

JOSEPH PETSCHE

SAN JOSE STATE UNIVERSITY

 cognella® ACADEMIC PUBLISHING

Bassim Hamadeh, CEO and Publisher
Kassie Graves, Director of Acquisitions and Sales
Jamie Giganti, Senior Managing Editor
Miguel Macias, Senior Graphic Designer
Carrie Montoya, Manager, Revisions and Author Care
Natalie Lakosil, Licensing Manager
Kaela Martin, Project Editor
Abbey Hastings, Associate Production Editor

Cover image copyright © 2014 Depositphotos/homeworks255.

Printed in the United States of America

ISBN: 978-1-5165-1306-2 (pbk) / 978-1-5165-1307-9 (br)

CONTENTS

PREFACE

I want to thank Cognella Academic Publishing for this opportunity to not only design a textbook for classes like those I teach at San Jose State University, but to fulfill my dreams of creating a book of my own, in the way that I wanted to write and design it. This creative freedom allowed us to produce a textbook that stands out among its competitors because of its lower price and focus on a greater variety of prehistoric organisms, which is what books like these are supposed to be all about.

Another point worth noting here is that this book describes our current understanding of the known fossil record, which is incredibly rich, but at the same time extremely limited in scope. Consider that most organisms that ever lived never left behind any discernible fossils, and of those that did, there is a bias in the fossil record towards hard tissues such as bones, teeth, and shells. Now, consider that our access to fossil-bearing rocks is extremely limited, and when we do get access, budget and time constraints limit our exploration to the upper few feet of the crust. Thus, the part of the fossil record that has been observed so far must constitute only a fraction of a percent of all the fossils that exist. It's safe to say that we are in store for many profound discoveries in the future that will add more detail to our understanding of the history of life, and possibly upend long-standing theories we tend to take for granted nowadays.

Despite these limitations, the fossil record provides overwhelming evidence that evolution has influenced the succession of life since it first began, nearly 4 billion years ago. Layers of rock can be read like a book, and the story we see is a predictable pattern of increasing complexity and diversity. We are able to discern this pattern because of the contributions from generations of scientists who worked relentlessly to discover, describe, and catalogue rocks and fossils. Evolution is a theory based on observation; it is empirical and backed by solid, physical evidence. Yet, we must be careful not to turn evolution into dogma. Well-established theories such as evolution have the tendency to become what some people consider "fact." When a theory is treated as fact, people stop questioning it, and instead develop a type of "faith" in the theory. Once people become too comfortable with a theory they tend to stop questioning it, and if that happens, where is the impetus for further scientific study?

ACKNOWLEDGMENTS

First and foremost, I want to thank my wonderful wife, Sarah, for her patience and support while I toiled away for 16 months writing this book. I also want to thank my mom, dad and my brother Steve for their support of my interests in rocks and fossils as a kid. I still remember the excitement of "discovering" my first trilobite in my Christmas stocking when I was eight.

I would also like to thank Niles Eldredge for writing one of the books that inspired me as a teenager: *Fossils, the Evolution and Extinction of Species*. Dr. Eldredge was also kind enough to respond to my questions and provide me with new insights while writing this book. I also thank the Paleontology Research Institute (PRI) in Ithaca, NY, and the Estrella Mountain Regional Park (AZ) visitors' center for their help. In addition, I would like to thank the San Jose State University Department of Geology for allowing me to do what I love, teach science.

This textbook is designed to be more streamlined and affordable than other books in the genre. Textbook prices are, in part, based on the licensing of photos and diagrams, which in many cases cost money and raise the price of the book. In an attempt to keep this book affordable, I used as many duty-free images as I could, many of them from my own collection, and many others from Wikimedia Commons. I want to thank all the people who generously uploaded photos and other media to be used free by the public; their wonderful donations helped keep the price of this book down.

This book is dedicated to my wife, Sarah.

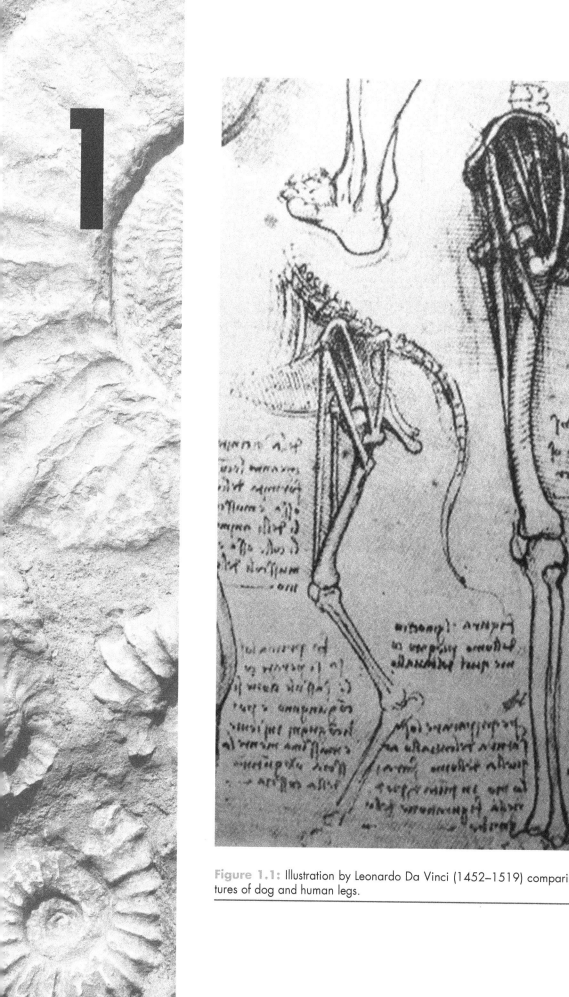

Figure 1.1: Illustration by Leonardo Da Vinci (1452–1519) comparing bone structures of dog and human legs.

SCIENCE

SCIENCE AND THE SCIENTIFIC METHOD

In November of 2005, the Kansas State Board of Education, while updating its educational science standards, literally redefined **science** for its state. The traditional, long-standing definition of science was along the lines of *the human activity of seeking a natural explanation for things that we observe.*

State lawmakers and the lobbyists that influence them are generally not scientists. Despite this lack of qualifications, the board was able to create a new definition of science: "a systematic method of continuing investigation that uses observation, hypothesis testing, measurement, experimentation, logical argument, and theory building to lead to more adequate explanations of natural phenomena" (Kansas State Board of Education). This new definition, although seemingly benign, gave Kansas the right to treat creationism as a theory equal in scientific merit as that of evolution. By removing the words "natural explanation," supernatural explanations were now invited into the science classroom. Replacing the word "natural" with "logical" in the definition invited research to operate within the framework of religious logic. For example, it could be scientifically stated that creationism is logical, given the complexity of life today. Two years later, however, this change was voted down.

Although the concept of science is idealized as objective, the *practice* of science cannot avoid the involvement of the human brain, which is arguably incapable of purely unbiased thoughts. For example, one of the commonalities between science and religion is *dogma*. Some scientific theories are so well established that many people, including scientists, accept them as fact and don't give them much critical thought. Widespread acceptance doesn't make a theory any less valid, but it may bias the analysis and interpretation of data and cause an apathy that impedes the search for alternative theories.

Bias, luckily, has a limited effect on science. Supernatural phenomena are by nature irregular and do not follow any known rules, which makes them untestable. Because the results of scientific experiments are subject to scientific scrutiny and can be retested, results affected by any form of bias, be it scientific or religious, can be discovered and weeded out. Religious explanations cannot be scientific because they ignore or do not conform to the observed physical rules that govern our universe; they are only responsible for adhering to their own inner logic.

People who actively attempt to inhibit the objective investigation of natural phenomena have historically been motivated by religion. For instance, Rhazes (865–925), a medical researcher in Baghdad, created a compilation of empirical knowledge about medicine in his book *Continens Liber*, which was later used to literally beat him blind by a Muslim priest who was offended by his Western viewpoint.

Galileo Galilei (1564–1642) was placed under house arrest after he was tried and convicted for heresy in 1633 after he published theories supporting Copernicus' heliocentric version of the solar system (Figure 1.2). He remained under house arrest until he died twelve years later. Galileo had it easy compared to Giordano Bruno (1548–1600), who in 1600 was burned alive at the stake for his support of the Copernican viewpoint.

Today, the main enemy of science is ignorance, apathy, and a general lack of critical thought. For example, a series of minor earthquakes that hit L'Aquila, Italy, in the spring of 2009 prompted a team of geoscientists to meet to assess the quakes and the potential of future seismicity.

The conclusions of the meeting stated that "a major earthquake in the area is unlikely but cannot be ruled out." Furthermore, Enzo Boschi of the Italian Civil Protection Agency said that "because L'Aquila is in a high-risk zone it is impossible to say with certainty that there will be no large earthquake."

The ambiguity of these conclusions reflects our inability to predict earthquakes, an indication of the progress that still needs to be made in the field of seismicity. No scientific method exists to predict earthquakes. Furthermore, these Italian scientists did not, by any stretch of the imagination, indicate that Italy was safe from future tremors. Unfortunately, a government official who had no academic experience with seismology claimed on national media that there wasn't much of a danger of a larger quake because enough energy had been discharged.

On April 6, 2009, a magnitude 6.3 earthquake struck L'Aquila, killing nearly three hundred people (Figure 1.3). As a result, the six Italian scientists who claimed "it is impossible to say with certainty that there will be no large earthquake" were convicted of manslaughter and sentenced to six years in jail for failing to predict the quake, which any geologist knows is *impossible to do thus far!*

History clearly shows that the pursuit of knowledge has been discouraged by threats of imprisonment or death. The concepts discussed in this book are losing ground in the public schools today. States such as Kansas, Ohio, Kentucky, Colorado, New York, Missouri, Oklahoma, and Montana have already altered or attempted to alter their science standards to include creationism as a science and/or to shed a more critical light on evolutionary

Figure 1.2: Page from Galileo's notebook showing the four largest moons of Jupiter, with a modern illustration for comparison. This observation helped us realize that not all things orbit the Earth.

Figure 1.3: President Barack Obama tours an area affected by the earthquake in L'Aquila, Italy, July 8, 2009.

theory. The latter is acceptable, as scientific theories *should* be scrutinized. The inclusion of creationism, however, serves to weaken science by infiltrating a practice that strives to be objective with ideas that have no real basis in the observable natural world. This is why the scientific method (Figure 1.4) is so important; it sets a standardized framework for collecting, analyzing, and interpreting data that is designed to reduce the potential of misuse and dishonesty.

FIELDS OF SCIENCE THAT CONTRIBUTE TO THE STUDY OF EARTH'S HISTORY

The study of prehistoric life is multidisciplinary; several fields of science converge with a powerful sort of synergy induced by the collaboration of ideas.

Geology is a field of science that studies the physical structure and materials of the earth, how and when they formed, and the processes that act on them. It provides the deep-time framework in which evolution can operate. Through relative and absolute dating methods, geologists determine the ages of fossils and the rocks they formed in (Figure 1.5). The study of rocks and minerals and their formation allows us to determine the variety of ways in which fossils form and the best conditions to preserve them. By correlating strata, a much more comprehensive understanding of the history of life has been stitched together from stacks of sedimentary beds scattered across the planet (Figure 1.6). The study of plate tectonics has allowed geologists to

The Scientific Method

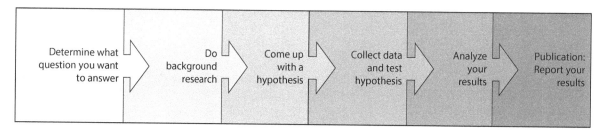

Figure 1.4: The steps of the Scientific Method.

explain the identical fossils of organisms on continents now separated by oceans. Geology provides us with the understanding of how earthquakes, volcanoes, mountain building, asteroid impacts, and a multitude of other phenomena have influenced the evolution, adaptation, and extinction of organisms over time. Finally, geologists help us understand the conditions present in the early Earth when the first living organisms are thought to have originated.

Biology is the study of living organisms. Without biology, this book would not exist, nor possibly you, as biologists provide us with the ability to cure disease, heal injury, and fix eyesight. Evolutionary theory requires the knowledge that only the careful study of organisms and their behavior, anatomy, and physiology can provide. Biologists started the conversation of evolution and are an integral part of its study.

Figure 1.5: This Pliocene-age formation at Capitola Beach, CA, was deposited on the sandy seabed close to shore several million years ago. The layer of broken shells may have been deposited during a large storm.

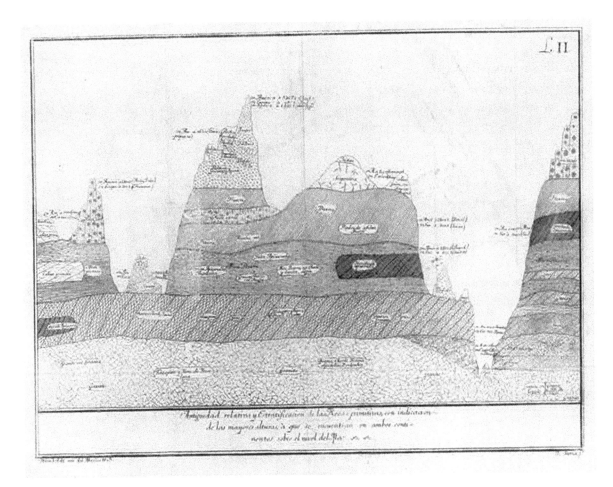

Figure 1.6: Sketch of volcanic rock strata in Mexico by Alexander Von Humboldt, 1803.

Paleontology is the study of fossil organisms and the physical evidence they leave behind. Paleontologists unearth, describe, and chronicle the fossil record and provide the physical evidence of evolution through time. They belong to the most important field of science contributing to our understanding of prehistoric life. Moreover, most of the history of our own species is provided to us by paleontologists.

Archaeologists and **Anthropologists** study the organic and cultural remains of past humans. These fields help us understand human history and evolution by resolving the chronology of the physical and cultural development of our ancestors through the study of their bones and artifacts.

Meteorology is the study of weather and climate. The fossil record contains evidence of past climate change that can only be understood with a working knowledge of today's climates and the effects they have on the crust and the organisms that live on it. The evolutionary pathways of life throughout Earth's history are greatly altered and periodically truncated by major climatic events such as ice ages and periods of global warming much more extreme than we are currently witnessing.

Oceanography is the study of oceans and the natural processes that affect them. The majority of fossils we find today are of marine organisms, and to understand their ecological niche during their existence, we need to understand the complex interplay of organisms with each other and their environments in oceans today.

THE FOSSIL RECORD

The **fossil record** is a natural database containing physical, measurable information about past life. It is organized chronologically within stacks of sedimentary beds like the pages of a book. Typically, when we refer to the "fossil record," we mean the record of fossils that has been uncovered thus far. It should be noted, however, that the vast amount of fossils we've collected constitutes a tiny fraction of what still remains locked away in our crust. Fossils are generally found only when they are weathering directly out of rocky outcrops or are accidentally uncovered during excavation of shallow pits and quarries.

The primary reason why our understanding of prehistoric life is extremely limited is the inability of scientists to do anything more than merely scratch the surface of our planet. Most fossil-bearing rocks are presumably underwater, on private property, in regions where field research is prohibited, or covered in ice. If we are lucky enough to get access to these rocks, time and budgetary constraints prevent us from digging too deeply. One can only imagine the plethora of yet-unknown organisms that lie in rocks just beyond our reach.

Moreover, most organisms never fossilize; they decay or are eaten instead. The known fossil record represents a minuscule portion of all the fossils out there, which themselves are represented mostly by hard tissues such as bones, teeth, and shells.

Therefore, there is great potential for future discoveries that may cause us to fundamentally rethink evolution and the origins of life. For now, we can only interpret the body of evidence discovered thus far, and this book summarizes our current understanding of this very long, complex, and poorly understood history.

THE AGE OF THE EARTH

The age of the Earth has historically been very difficult to pin down. In the Western world, the age has traditionally been determined by theologians who simply counted the generations in the Bible, adding up all the years it would have taken for them to live, which is roughly six thousand years. Recent Gallup Polls[1] have shown that today, approximately 30 percent of adults in the United States interpret the Bible literally and more or less adhere to the Young Earth Theory.

The Young Earth Theory is not supported by scientific evidence. The study of geological processes suggests that much longer spans of time are required to allow mountains to rise, layers of rock to form, and erosion to shape the surface of the Earth. In terms of time-scales of geological processes, creationism is founded on the notion of **catastrophism,** which states that geological processes have been rapid and extreme throughout Earth's short existence. Mountains, ocean basins, volcanoes, sedimentary layers, etc., are thought to have formed in very short spans of time and are attributed not to the processes we observe today but rather divine intervention by a god. Despite the longstanding influence of catastrophism and, currently, the over seventy million people in the United States alone who believe in it, geology and related fields have been able to provide overwhelming scientific evidence that points to the contrary; the Earth is billions of years old, not thousands. The long history of geological time is referred to as deep time.

Prior to the development of modern dating techniques about one hundred years ago, the counterargument to the Young Earth Theory was based primarily on **uniformitarianism,** the assumption that the rates at which geological processes occur today are the same as they have always been. Since most processes observed in nature today

---

1   Conducted via phone over May 5–8, 2011.

Figure 1.7: A) James Hutton (1726–1797) is informally known as the "father" of modern geology. B) Nicolas Steno (1638–1686) provided criteria for establishing relative ages.

Figure 1.8: Aftermath of a tsunami in Japan, March 11, 2011; an example of a catastrophic event that punctuated the generally "uniform" history of slow-going geological processes in the area.

Figure 1.9: Siccar Point, a classic and well-known angular unconformity that helped James Hutton visualize the incredible amount of time it would have taken for erosion and deposition to create such a complicated juxtaposition of layers.

happen at very slow rates, so did these processes in the past. This implies a much older Earth than creationism. The primary tenet of this assumption is that the present is the key to the past. This notion, which is supported by many lines of scientific evidence, is the foundation of modern geology.

Uniformitarianism and deep time were developed in a scientific context by James Hutton (1726–1797). Known as the "father of modern geology," Hutton (Figure 1.7A) was an astute naturalist, physician, and geologist based out of Edinburgh during the 1780s. He published articles outlining evidence that supports deep time as a framework for studying the history of our planet.

One caveat is that we do, in fact, recognize that there are some processes in nature that are catastrophic, such as tsunamis, volcanic eruptions, earthquakes, landslides, and asteroid impacts (Figure 1.8). These catastrophic events punctuate long periods of relative quiescence, during which processes shaping the Earth are generally slow and steady.

Charles Lyell

Figure 1.10: Charles Lyell.

Nicolas Steno (1638–1686) (Figure 1.7B) was an early worker in geology during a time when the field was becoming a mainstream science. He posited several principles of stratigraphy, which are laws that help us interpret the relative ages of various rocks. These laws, as well as Hutton's conclusions, were later elaborated on in the book *Principles of Geology*, written by Charles Lyell, a British geologist, in 1830 (Figure 1.10). This set of principles is now established as the Laws of Relative Dating. (*Principles of Geology* was the only book Darwin brought on his five-year expedition on the HMS *Beagle*.)

2

Figure 2.1: A skeleton from a male mastodon that roamed upstate New York about 15,000 years ago, Museum of the Earth, Ithaca, NY.

FOSSILS AND THE GEOLOGIC TIMESCALE

WEATHERING AND EROSION

It would be very difficult to find a natural setting on which fragments of rocks and minerals, called **sediments**, aren't scattered all over the place. Sediments can be found in various sizes ranging from clay all the way to large boulders, and in layers ranging from a thin dusting to accumulations miles deep. A large variety of **sedimentary structures** are also formed when sediments accumulate in specific ways. All loose rocks, excluding meteorites, were at one time or another attached to the crust. Furthermore, there are a lot of dissolved minerals in streams, lakes, oceans, and groundwater. **Weathering** is a collective term that refers to any process that strips material away from the solid crust. **Physical weathering** (Figure 2.2) includes processes that mechanically break rocks into sediments. **Chemical weathering** involves chemical reactions that cause rocks and minerals to disintegrate, disaggregate, and/or dissolve.

PHYSICAL WEATHERING

Physical weathering results in the formation of sediments, which can later become cemented or compacted together into **clastic sedimentary** rocks. **Abrasion** is a common type of weathering and involves the grinding or shattering of rocks due to collisions with other rocks and the solid ground. There are many ways rock can weather into sediments.

Glaciers are powerful weathering agents. Countless rocks are embedded within these moving masses of ice, especially along the bottom. These rocks drag across and scour the ground, in some cases grinding the bedrock into a fine powder. Most rocks in a glacier are weathered from bedrock farther upstream and become the agents of weathering themselves. In places, bedrock weathered by glaciers can be incredibly smooth, displaying **glacial polish** or even linear scrape marks called **striations** that indicate the flow direction of the ice.

Waves pound along shorelines where rocks are exposed, especially narrow beaches with steep cliffs. Not only does the collision of rocks within the water and along the beach wear them down, the water itself can result in

Figure 2.2: Yosemite Valley provides "textbook" examples of various types of physical weathering. Pleistocene glaciers carved this valley out of granite, the frozen remains of Cretaceous magma chambers.

cliff retreat. Waves scrape away material from the bottom of the cliff, creating an unstable overhang that eventually crumbles onto the beach below. For example, many houses perched along the California coastline have been lost due to cliff retreat over the years. Tiny fragments of these houses are now scattered for miles along the coastline.

In rivers, rocks tumble along the bed, collide with other rocks, and are blasted by sand, silt, and pebbles. Rocks that first enter rivers tend to be fairly angular in shape. The longer these rocks are exposed to this abrasive environment, the smoother they typically become (Figure 2.3). The corners and edges of the rocks (asperities) chip away easily, but the flatter surfaces are more resistant.

In places where the air becomes laden with wind-blown sand and silt, the surfaces of exposed rocks are sand-blasted. Smaller rocks on the ground will weather most in the upwind direction and can become smooth or faceted on one side, or several if the wind shifts. We call these rocks **ventifacts**. With the exception of large dust storms, the wind carries sand no more than a few meters off the ground. Outcrops of rock that are taller than this can weather more along their base than their tops and can become mushroom-shaped over time (Figure 2.4).

In addition to abrasion, which is typical in most settings, other types of physical weathering are more limited in extent.

Many rocks will weather into flakes or sheets, partly due to their internal structure. For example, in climates such as those found in mountainous areas, granite weathers into slabs ranging in thickness from several centimeters to several meters. This occurs because granite forms deep in the crust and has therefore moved to a new environment of lower pressure if exposed at the surface. Fractures due to pressure release form parallel to the surface of the exposed granite, and these sheets of granite between the fractures break apart even further when exposed to other weathering processes (Figures 2.5 and 2.6). Intricate networks of fractures commonly form in any mass of rock

Figure 2.3: A) Cobbles along a creek bed are weathered into rounded shapes; B) an angular cobble may become C) rounded if given time.

that had experienced a reduction of pressure. **Unloading** is the stripping away of overlying rock due to erosion or faulting, resulting in the displacement of deep crustal rocks to the surface. Large-scale fractures in rock are called **joints** and typically form perpendicular to tension stresses in the crust. For instance, pressure-release fractures in the granite mentioned above are referred to as **exfoliation joints**. Tension can commonly result from tectonic stresses as well as the expansion of rock that accompanies the lowering of pressure during unloading.

Figure 2.4: Wind erosion has weathered the bottom of this rocky outcrop in Eduardo Avaroa Andean Fauna National Reserve, Bolivia.

Figure 2.5: A) One of the "pinnacles" at Pinnacles National Park, CA, provides a great face for rock climbers to scale. Sets of diagonal joints cross each other on this outcrop of rhyolite. B) Water and sediments spill over this cliff made of 400-million-year-old shale near Ithaca, NY. C) A rock slide in Yosemite Valley, CA, triggered by the sudden detachment of a granite sheet.

Figure 2.6: Roots of this Jeffrey pine have grown horizontally within a pressure-release fracture, breaking the granite above it (left). Hematite, a mineral formed by the oxidation of iron, colors this sandstone in Arizona red.

Living organisms are the cause of a surprising amount of weathering. **Organic weathering** is a term for a variety of interactions between living things and the crust they live on. This type of weathering is all around you. You will likely have noticed sidewalks that have been buckled by tree roots in your neighborhood. The expansion of root systems from many types of plants can be forceful enough to pry rocks apart and widen existing fractures (Figure 2.7).

Figure 2.7: Roots from this former tree had grown straight through this outcrop of chert in the Marine Headlands of California.

Along shorelines, rocks become covered in bivalves such as clams, which include species that have developed the ability to bore into solid rock. By attaching to a rocky surface, then scraping the rock with the back edges of their shells, some clams can burrow themselves to a safe depth. This enables them to remain attached despite pounding waves and frequent visits by predators.

Humans alone account for tremendous amounts of physical weathering (Figure 2.8). Each year, mining activities account for tens of billions of tons of rock removed from the crust. Every tile, statue, countertop, headstone, and building facade made of rock was at one time part of the solid Earth. It is even likely that the spot on which you sit right now lies directly above rock or sediments that were somehow modified by construction or landscaping, and possibly earlier by agricultural practices.

Frost wedging is a form of physical weathering that affects rock in climates where the temperature drops below freezing. Unlike most other substances that melt and freeze, water takes up more space as ice than it does in liquid form; this is why ice floats. Cracks in bedrock may collect water from rain, streams, and even groundwater. When the water freezes, it expands, further prying the fracture open (Figure 2.9). Piles of angular rocks called **talus** form at the bases of rocky cliffs due to frost wedging.

Figure 2.8: Human-induced organic weathering.

Thermal contraction and expansion occur when materials, such as rock, grow and shrink repeatedly. The daily solar heating of rocks at the surface causes them to expand and subsequently shrink once the sun sets. Since the rock is heating and cooling most at its surface, the amount of expansion and contraction decreases deeper in the rock. This creates an uneven distribution of pressure and the formation of fractures in the rock. A related type of weathering affects certain types of clay that swell when they absorb water and shrink when they dry.

Physical weathering results in the formation of most sediments, but not all. For example, **clay**, the smallest sediment size, typically doesn't form due to the chipping away of rocks like other sediments. Instead, clay is actually a type of mineral produced during a different type of weathering altogether.

CHEMICAL WEATHERING

Chemical weathering results in chemical changes of the rock and doesn't directly result in the formation of sediments, although it can certainly aid mechanical weathering. Newly exposed surfaces of rocks, such as those created during mechanical weathering, are most vulnerable to chemical weathering. In many cases, chemical weathering results in the formation of new minerals from reactions that destroy old minerals. Although the minerals formed during chemical weathering are generally more stable, the reaction itself can loosen the inter-crystalline fabric of a rock, allowing it to disaggregate easier.

Hydrolysis is a type of chemical weathering that results in the formation of clay minerals. Feldspar, the most common mineral in the crust, undergoes hydrolysis when exposed to water. During hydrolysis, free hydrogen (H+) and hydroxide (OH)- ions in water take the place of other ions in the feldspar crystal. The feldspar changes into flat, microscopic clay particles that flake off easily. On average, igneous rocks are about 60 percent feldspar. Granite, for example, will disaggregate when the feldspar, a formally solid part of the rock, changes into soft clay, allowing other loosened mineral crystals to spall off the surface of the rock. The abundance of clay on the Earth's surface is a testament to the amount of feldspar present in rocks.

Oxidation is the formation of "rust" at the expense of metals such as iron that are present in silicate minerals. For instance, reddish-colored rock and soil are usually attributed to oxidized iron, also called hematite.

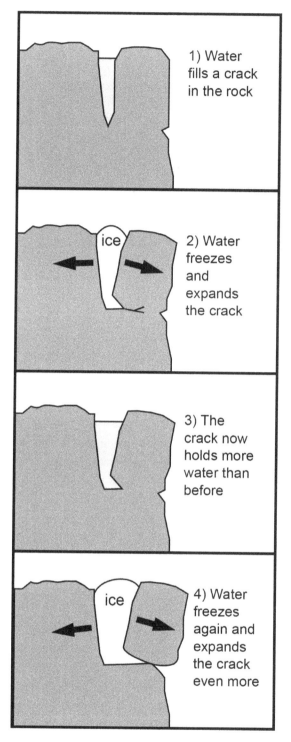

Figure 2.9: Stages of frost wedging.

1) Water fills a crack in the rock

2) Water freezes and expands the crack

3) The crack now holds more water than before

4) Water freezes again and expands the crack even more

Figure 2.10: Limestone weathered by scouring and dissolution forms a wall in Mosaic Canyon, Death Valley, CA.

Dissolution is the process by which minerals dissolve in water. In this case, no fragments of rock or new minerals are produced. The mineral simply dissolves; thus acidic water plays an important role in this type of chemical weathering. Water absorbs carbon dioxide, and in the process, carbonic acid $(H2CO_3)^-$ is formed. Although this is not the only way water becomes naturally acidic, it is an important reaction because minerals such as calcite $(CaCO_3)$, a carbonate mineral, dissolves easily with a little carbonic acid in the water (Figure 2.10). Water carrying these dissolved ions may redeposit the mineral as a coating on the surfaces of sediments, cementing them together in the process. In areas of abundant carbonate rock such as limestone and marble, calcite may reprecipitate as stalactites and stalagmites in caverns weathered out of the rock.

Over large areas of the crust where limestone and/or marble forms the bedrock, the collapse of underground caverns that had been weathered out of the rock due to dissolution can cause the entire landscape to have an appearance unique to these rock types. Known as **karst topography**, these landforms caused by dissolution are most typically found as sinkholes that dot the landscape and are connected by a complex and crumbling network of subterranean caverns (Figure 2.11).

Figure 2.11: This satellite image of Florida shows karst topography in the form of numerous water-filled sinkholes. Much of Florida's bedrock is limestone.

All of the examples above are considered weathering processes and could not be considered erosion without an additional step. **Erosion** is the *transport* of weathered products to a new location. In other words, erosion involves two processes: the weathering of rocks and the removal of the weathered material via some current of water or air, or by gravity.

SEDIMENTARY ROCKS

The processes that turn loose sediments or dissolved ions back into solid rock again are collectively referred to as **lithification**. There are several common ways in which sedimentary rocks can form (also see Figure 2.13):

PRECIPITATION

Certain minerals, such as calcite, are prone to dissolution weathering and end up as ions dissolved in groundwater. When the concentrations of these minerals get too high, they can reprecipitate, or grow back into solid crystals again. This is easily observed in limestone caverns, for instance, where thin films of calcite crystals left behind by dripping groundwater accumulate and slowly form crystalline structures such as stalactites and stalagmites. **Limestone**, a sedimentary rock in which caverns form, is itself formed by the deposition of calcite crystals grown within ocean water.

Lithification is the process by which loose sediments are stuck together to make a solid clastic sedimentary rock. There are two general ways this takes place:

CEMENTATION

In a layer of clastic sediments, groundwater can deposit mineral coatings on the sediments and eventually "glue" or cement them together (Figure 2.12). The growth of minerals such as quartz, calcite, and hematite within the spaces between sediments is a common way that clastic sedimentary rocks are created. The strongest of these cements is quartz, which tends to be the mineral associated with sedimentary rocks that are relatively more resistant to weathering.

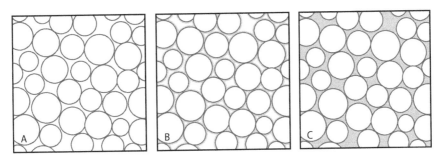

Figure 2.12: A) Loose sediments saturated with water; B) Water begins to deposit crystals of a cementing mineral on the surfaces of the grains. The grains begin to stick together. C) The grains have been completely cemented together; the pore spaces are all filled with cement.

COMPACTION

Another way sediments can become lithified is when sedimentary grains are squeezed together tightly enough to make them stick. Typically, **compaction** happens with clay particles, which have a slight electrical charge and a flat shape, causing them to stick together in a pattern not unlike the shingles of a roof.

CLASSIFYING SEDIMENTARY ROCKS

Sedimentary rocks can be divided into two major groups based on the types of weathering their components are associated with (Figure 2.13). **Clastic** sedimentary rocks are formed by the **cementation** or compaction of pre-existing sediment particles. **Chemical** sedimentary rocks form when dissolved minerals precipitate back into solid crystals, although when this happens during cementation, we still call it clastic. When fossils are present in either of these rocks, we attach the prefix *bio-* to the name (i.e., *bioclastic*). When mineral deposits are left behind when bodies of water such as lakes dry up, they are called **evaporites**. Salt, also called halite, is a very common evaporite.

Clastic sedimentary rocks are categorized based on the sizes and shapes of the sediments they are composed of. Two measurements are of particular importance when interpreting the environment in which the rock had formed:

Rounding is the measurement of how smooth a rock has become due to abrasion-related weathering. Typically, rocks exposed to abrasion, especially those in rivers and along shorelines, begin as **angular** rocks and become more rounded as time progresses.

Size Designation		Diameter	Type of Clastic Rock it Forms	Typical Depositional Environment
Fine Grained ↕ Coarse Grained	CLAY	less than 1/256 mm	Mudstone (Shale)	Typically at the bottom of bodies of water
	SILT	1/256m - 1/16 mm	Siltstone	In shallow water, especially streams
	SAND	1/16 - 2 mm	Sandstone	Beaches, deserts, and rivers
	PEBBLES (GRAVEL)	2 mm - 6.4 cm	Conglomerate and Breccia (when mixed with smaller-sized grains)	Rivers, rockslides, rocky beaches
	COBBLES	6.4 - 25 cm		
	BOULDERS	25 cm and greater		

Figure 2.13: Sizes of sedimentary grains, the rocks they form, and a few examples of the typical settings they form in.

Sorting describes the relative range of grain sizes in a sedimentary rock. A *poorly-sorted* rock, such as **conglomerate**, contains a mixture of different grain sizes. A *well-sorted rock*, like **sandstone**, contains mostly one grain size (Figures 2.14 and 2.15).

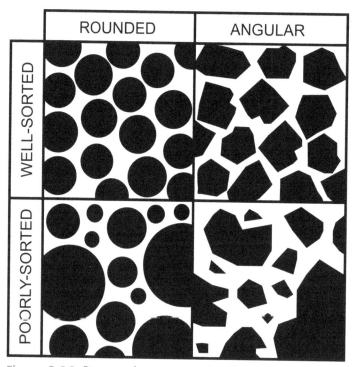

Figure 2.14: Diagram showing examples of sediment shapes and size ranges.

Figure 2.15: A close-up view of a mature, quartz-rich sand from the Netherlands.

DEPOSITIONAL ENVIRONMENTS

Sedimentary rocks give us clues about how and where they formed. Measuring the grain size, sorting, and shape of the sediments in a clastic sedimentary rock, for example, can help us understand what the environment looked like when these sediments were first deposited.

Sediments typically pile up in places where the currents carrying them, such as rivers or wind, lose energy. The overall size and shape of sediments indicates the level of kinetic energy inherent in the current carrying them. For instance, deep, fast-moving rivers are capable of carrying large boulders and making them round along the way. Strong sustained winds can move sand around, but not big boulders. Only mud-sized particles tend to make it to the deeper parts of the oceans and lakes.

The maximum size of sediments a river can carry is referred to as the river's **competence**. The total volume of sediments being moved by a river over a given period of time is called **capacity**. The more energetic a river is, the greater its competence and capacity tend to be.

COMMON SEDIMENTARY ROCKS

FINE-GRAINED CLASTIC ROCKS

Sediments such as clay and silt can only settle to the bottom of relatively calm water. Although energetic rivers can easily carry these tiny particles, the turbulence of the water makes it very difficult for fine sediments to accumulate on the river

Figure 2.16: Desert pavement in Death Valley National Monument, CA. Camera lens (52 mm) for scale.

bed. Mud and clay that does accumulate in rivers tends to do so along the banks, especially the inner shorelines of meanders, where the river's flow velocity is reduced.

Sediments carried by the river will eventually settle once the river empties into a lake or ocean. At the point where the water leaves the stream channel, the kinetic energy of the water attenuates, slowing it down and making it easier for finer-grained particles to settle to the bottom of the water. Compared to larger-sized grains, the mud and clay will be able to make it farthest out into lakes and oceans because they can take longer to settle.

SANDSTONE

Sandstone is usually very well-sorted and can form in a variety of environments. Sand isn't just a grain size; it's what most sediments become if given the opportunity. In that sense, sand is considered to be "mature," which means it represents the endpoint of most types of mechanical weathering.

Rivers carry tremendous amounts of sediments from land and dump them into standing bodies of water such as lakes and oceans. Each year, the world's rivers strip nearly ten million tons of sediments off the land and carry them to larger bodies of water; the number is nearly twenty-five million tons if human-derived sediments are included. Along the way, these sediments are weathered into progressively smaller particles of rock until they become sand and silt. It becomes a lot harder for sediments the size of sand to break apart further; the small mass of a sand grain gives it only a tiny bit of kinetic energy, making it difficult to accumulate enough energy to fracture the grain. Sand can definitely become smaller, however, especially if it is ground between larger rocks. Where the bottoms of glaciers scrape the bedrock, or where rocks are crushed under landslides, for example, sand can easily be broken into silt or even clay-sized grains. The light blue color of glacial lakes is attributed to the presence of *rock flour* in the water. Rock flour, also called *glacial flour*, forms underneath the glacier by the grinding of rocks into a silt-sized powder.

A mature sand, in the context of rivers, is something produced only after the sediments have been exposed to the abrasive environment of a river for protracted periods of time and/or carried long distances. It also means that minerals less resistant to chemical weathering, such as feldspar, have been removed from the sand, making it especially rich in quartz.

In the context of **aeolian** (wind-blown) sand, the maturity is based on how well the wind has sorted the sediments. Wind is typically strong enough to lift silt and clay from the ground, leaving behind sand and larger grains. This process is called **saltation**, and in places like the desert, where sediments on the ground are often poorly sorted, saltation can result in an accumulation of larger grains, especially pebbles, at the surface. This concentration of large sediments is called **desert pavement** (Figure 2.16).

CONGLOMERATE

Unlike breccia, conglomerate is composed of rounded grains, implying a depositional environment within or near flowing water. Typical settings of conglomerate deposition include river beds, shorelines weathered by wave action, and glaciers. Unlike wind and water, flowing ice doesn't sort sediments. The semi-rigid structure of glacial ice is capable of carrying sediments of nearly every size, and when the ice eventually melts, the

Figure 2.17: Yeager Rock, a glacial erratic, as exposed in a road cut in Washington State.

sediments simply pile up on the ground with no particular layering or internal arrangement. Glacial sediments, called **till,** are some of the most poorly sorted sediments that exist besides those found in rockslide deposits. Till can include extremely large boulders called **erratics** that can be found in precarious spots (Figure 2.17).

BRECCIA

Given the angular shape of its sediments, it can be inferred that breccia represents the deposition of rocks that haven't traveled very far from where they were weathered. **Breccia** forms most commonly from rockslide deposits and talus piles at the bottoms of cliffs.

Another source of breccia are faults in the brittle upper crust. Rocks along and within faults have the tendency to shatter and crush each time the fault moves during an earthquake. In some extreme cases, the rock can be melted and refrozen into a thin sheet of glass after an earthquake.

LIMESTONE AND CHERT

Limestone is a chemical sedimentary rock composed of calcite, which is a calcium carbonate mineral. It commonly forms at the bottoms of warm, shallow seas and lakes. In these environments, which tend to be concentrated within 30 degrees of the equator, tiny crystals of calcite form within the water column and gently settle down to the bottom, where they form layers that later lithify into limestone. The calcite crystals may nucleate on free-floating particles in the water and precipitate inorganically, or the crystals can be produced by organisms in the water that grow calcitic shells or exoskeletons, such as certain plankton. Coral reefs are rich in carbonates and often form circular belts of limestone that ring volcanic islands.

Chert (Figure 2.18) forms in similar ways to limestone but is made of silica (SiO^4) rather than calcite. Some chert can form within beds of limestone while still underwater, from the concentration of silica into nodules that nucleate within the calcite. Chert also forms from the accumulation of diatoms and radiolaria, which are microscopic plankton that grow very complex, glassy silica exoskeletons.

Figure 2.18: 2.3-billion-year-old chert, Jasper Knob, Ishpeming, Michigan.

TRAVERTINE AND TUFA

Limestone may also form due to the evaporation of water in locations such as caverns and hot springs. **Travertine** is a chemical sedimentary rock made of calcite that forms when carbonate-rich water dripping from the ceilings of caves or over rocky ledges leaves behind thin films of calcite when it evaporates. This phenomenon is also responsible for the annoying chalky residues left behind in bathrooms by hard water, which is rich in dissolved minerals such as calcite (Figure 2.19B).

Tufa is another type of limestone that forms when calcite builds up along groundwater springs, especially where carbonate-rich water is pumped through lake sediments, cementing them together (Figure 2.20). One of the most famous tufa localities is Mono Lake, California, an alkaline body of water over a million years old that nearly disappeared due to water diversion in the 1970s and 1980s.

COQUINA

Some beaches and reefs, especially those nearer the equator, are constructed from thick accumulations of coral and shell fragments, sometimes with very few lithic sediments. Rock made from calcitic fragments of organisms cemented together by calcite is called **coquina** (Figure 2.21).

EVAPORITES

As the name implies, these chemical sedimentary rocks form when mineral-rich water evaporates, leaving behind a mineral crust (Figure 2.22). Evaporite minerals include halite (salt), calcite, and gypsum. The typical depositional

CLASTIC SEDIMENTARY ROCKS

SHALE

Shale and **siltstone** are composed of clay- and/or silt-sized grains. They typically form thin, parallel layers and are often inter-bedded with sandstone. These rocks usually form in wet, muddy environments, such as the ocean floor, riverbeds, dried-up lakes, and tidal flats.

Sandstone is made of cemented or compacted sand, ranges from fine to coarse in texture, and is typically well-sorted. Sandstones composed mainly of quartz grains are common, as well as types such as **graywacke**, which is composed of quartz and feldspar grains. Sandstone forms in sandy environments (of course), such as shorelines, desert basins, and along rivers and floodplains.

SANDSTONE

CONGLOMERATE

Conglomerate is a poorly-sorted sedimentary rock that is typically composed of pebbles, sand, and silt. This rock is similar to breccia (see below), but only has rounded grains. Conglomerate forms from water-worn sediments collected along rivers and coastlines.

Breccia is a poorly-sorted sedimentary rock. Unlike conglomerate, it's grains are angular. Breccia forms where freshly broken-rocks accumulate, such as in rockslides and within brittle fault zones.

BRECCIA

Figure 2.19: (A) Clastic Sedimentary Rocks.

CHEMICAL SEDIMENTARY ROCKS

LIMESTONE

Limestone is a chemical sedimentary rock that forms in a variety of settings, such as warm shallow seas, hot springs, caverns, lakes, and anywhere else where calcite accumulates in water to the point where it begins to form crystals. In the ocean depths, limestone forms from the accumulation of tiny calcite crystals and/or plankton exoskeletons.

Chert forms in similar ways as limestone in the oceans, except the mineral being precipitated is silica, a substance very similar to quartz. Biochemical chert is typically composed of radiolaria, which are planktonic microorganisms that extract silica from water to create tiny but astonishingly intricate exoskeletons.

CHERT

SPELEOTHEMS are rocks formed when minerals are precipitated underground caverns. Typically made of calcite or gypsum, these chemical sedimentary rocks form slowly when water leaves behind numerous layers of mineral coatings on the ceiling, walls, and floors of caverns.

CALCITE STALAGTITE including CROSS-SECTION

Evaporites are minerals that are deposited from solution, typically when water carrying dissolved ions evaporates, leaving behind mineral crystals. These types of rocks are common in deserts where bodies of water that accumulate dry up quickly, leaving behind playas composed of a mineral crust.

HALITE (salt) an EVAPORITE

Figure 2.19: (B) Clastic Sedimentary Rocks.

Figure 2.21: Coquina, a biochemical sedimentary rock composed almost entirely of shell fragments.

Figure 2.20: Layers of sandy lake sediments cemented together, a variety called tufa, Pyramid Lake, Paiute Territory, NV.

Figure 2.22: Pleistocene lake sediments rich in carbonate evaporites in Death Valley National Park, CA.

settings of evaporites are stagnant bodies of water in closed basins or along the shorelines of oceans. When the water evaporates, it doesn't take any of the minerals with it; as the water shrinks, the concentration of these dissolved minerals increases until crystals are forced to precipitate. Once the water is completely evaporated, a patch of mineral residue is left behind on the ground. These can be seen most easily in the desert, where they form dried lake beds called **playas**. This process is used by humans as a way to extract salt from seawater, wherein the water is diverted into evaporation ponds and the salt is harvested when the ponds dry up.

HOW FOSSILS FORM

Fossils are preserved remains and/or traces of organisms and provide a record of prehistoric life and evolutionary changes throughout geologic history. Our understanding of past life is somewhat biased, as some organisms fossilize better than others. Generally, soft-bodied organisms such as jellyfish aren't as easily fossilized as those with hard tissues such as bone, teeth, and shells.

There are several factors that favor the preservation of fossils.

ANOXIA

One of the most common effects that prevents fossilization is the decay of organic tissues. Water usually has oxygen dissolved in it, but some bodies of water, such as bogs and swampy ponds, are **anoxic**, meaning there is a low level of oxygen in the water. These environments tend to have an abundance of plant matter that falls into the water. Over time, the decay of the plants in the water removes much of its oxygen, so any new organic tissue that settles in the water tends to rot at such a slow rate that it can be preserved before decay has destroyed it. The accumulated organic matter, made up primarily of plant tissue, can become compacted and squeezed dry due to burial.

This layer, enriched in preserved organic matter, is at first called *peat*, but after further burial, pressurization, and breakdown of the organic molecules into simpler compounds, peat turns into *coal*. **Coal** is fossilized to a greater degree than peat, as it is more chemically altered and has lost an assortment of elements that had leached out into surrounding rocks.

Tar pits and amber can also envelop an organism quickly, removing it from the atmosphere and encasing it in an airtight medium.

DESICCATION

Desiccation is the process of drying out. Mummies are a good example of how tissues that dry out quickly can more easily fossilize because decay is less likely to occur. Desiccation also turns soft, moist tissue into hard, dry tissue, which is more rigid and will better withstand being crushed by overlying layers after burial.

FINE-GRAINED SEDIMENTS

Sediments are the packing material for many fossils, and the size of the sediments greatly affects the level of detail a fossil can be preserved with. Imagine that if you had to ship a delicate glass vase in a box of sediments, you likely wouldn't use large pebbles and cobbles to pack it in. Sand would encase the vase more evenly, and clay would be even better, making a tight, rigid seal around the delicate glass. In nature, fossils are preserved with the most detail when they are buried in very fine-grained sediments such as mud, silt, and sand. The ability of these fine particles to fit into small pores and crevices makes it more likely that fine details of the fossil will be preserved and not crushed.

RAPID BURIAL

Most fossils begin as dead organisms resting on the ground, and rapid burial makes it less likely that the remains will be eaten or scattered by scavengers. Rapid burial also reduces the chances that the tissue will decay to the point that it is eradicated from the fossil record.

This is true for fossils that form on land as well as fossils that form in layers at the bottoms of rivers, lakes, and oceans.

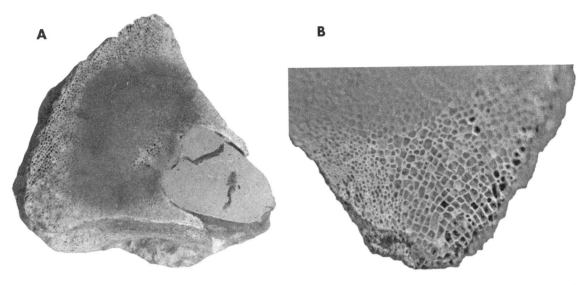

Figure 2.23: A) Permineralized whale vertebra; B) close-up showing some, but not all, pores filled with minerals.

HARD TISSUES

Tissues such as bones, teeth, shells, spines, and exoskeletons fossilize more easily than soft tissue. These tissues are often constructed in part by minerals such as calcite, quartz, and apatite, which are more chemically stable and less prone to decay than soft tissues are. Furthermore, the rigidity of these tissues makes it less likely they will be crushed by the weight of overlying layers of rock. Sharks, for instance, are one of the few cartilaginous fish still alive. Their teeth are easily fossilized, but the rest of their skeleton, made of softer tissue, tends not to be preserved. Much of what we understand about ancient sharks is based on the size, shape, and chemical makeup of their teeth.

TYPES OF FOSSILIZATION

There are two general types of fossils. **Body fossils** are the remains of the actual organism, whereas **trace fossils** represent evidence that indicates an organism was around, such as footprints, burrows, and preserved feces (**coprolites**).

BODY FOSSILS

Permineralization: Many fossils are formed when minerals from the surrounding rocks, usually carried by groundwater, grow within the pores and cavities of organic tissue. A permineralized bone, for example, has its porous marrow filled in with minerals, but some of the bone material is still there (Figure 2.23). Similarly, a permineralized shell will be filled in by new minerals while the shell itself is still mostly in its original state.

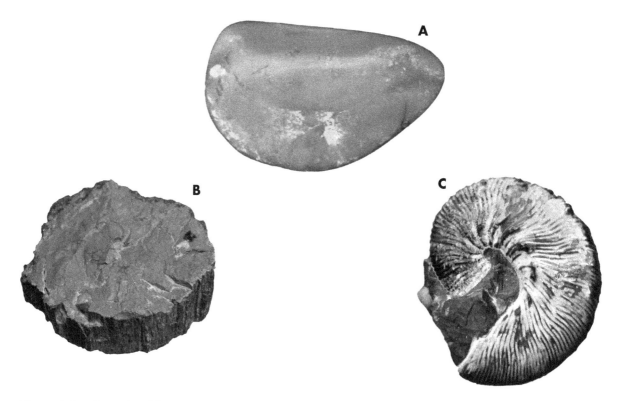

Figure 2.24: Examples of fossils replaced by opal (silica); A) clam; B) wood; C) ammonite, a marine cephalopod.

REPLACEMENT

After permineralization has filled in the pores and cavities within a fossil, the next step is for whatever remaining tissue there is to break down and and be replaced completely by new minerals (Figures 2.23, 2.24, 2.27). Quartz and calcite are two minerals that commonly permineralize and replace fossil tissue. Most dinosaur bones have been replaced completely by new minerals, making it nearly impossible to find any residual **DNA** in their remains. Still, there are some dinosaur fossils that, despite being very old, contain a tiny amount of soft tissue in the middle of the bone, remains of what used to be blood vessels. Perhaps the replacement of the outer surfaces of the fossils sealed in the soft tissue, protecting it for tens of millions of years.

RECRYSTALLIZATION

Not all fossils are replaced. Some tissues such as shells and exoskeletons are already made of minerals, and the only major change that occurs during fossilization is that the tiny, organically produced mineral crystals making up the tissue simply recrystallize without any replacement by new minerals. This is by far the most common with calcium carbonate and is seen primarily in fossils of shells that have been buried deeply enough to achieve the conditions necessary for recrystallization. In some cases, pressure isn't required for recrystallization at all; the dissolution and reprecipitation of the same carbonate material can achieve a similar effect. Recrystallization tends to fuse adjacent crystals together, creating larger crystals. This causes fine details in fossils to be destroyed, such as the original color and texture of shells, making them appear more granular and earthy. In some marbles, which result from the metamorphism of limestone, relict fossils of shells can still be seen as subtle outlines in the crystalline patchwork composing the rock.

CARBONIZATION

A thin film rich in carbon is usually what is left after a fossil has been carbonized (Figure 2.25). When fossils are flattened, or compressed, by pressure and the volatile compounds from the tissue have been dissolved and carried away by water, the relatively stable and immobile carbon gets left behind. Flat, small, and/or soft specimens such as leaves, insects, and jellyfish are the most common carbonized fossils, as they are compressed rather easily without much distortion. This process is also responsible for the formation of coal from peat.

Figure 2.25: Carbonized leaf in a gray mudstone.

Some types of body fossils still contain large amounts of original tissue. Sometimes referred to as **unaltered fossils**, these examples tend to be less common than the others since the circumstances required for such delicate fossilization aren't easily achieved.

MOLDS AND CASTS

If an organism buried in the sediments decays and eventually disappears, it may still leave behind a void shaped like the original organism. This void is called a **mold** and can later be filled in by sediments or minerals that take the shape of the mold, creating a cast. Simply put, a mold is an empty space that can create a **cast** if it is filled in (Figure 2.26). An ice cube is a cast from the ice cube tray, which is the mold.

In fact, ice is a good medium for preserving fossil tissue in more or less an unaltered state (Figure 2.28A). Many organisms, some extinct, have been found embedded in glacial ice in places like Siberia. Even human fossils, thousands of years old, have been uncovered in these environments with their flesh and organs still intact, albeit a little shriveled up and rotten. Of course, once removed from the ice, these frozen fossils don't last long unless preserved some other way by scientists. Mammoths and mastodons, for example, are found every few years trapped in bodies of ice which have not melted in thousands or even tens of thousands of years. Fossils such as these are rare and

Figure 2.26: Mold of a gastropod; the actual shell has weathered away.

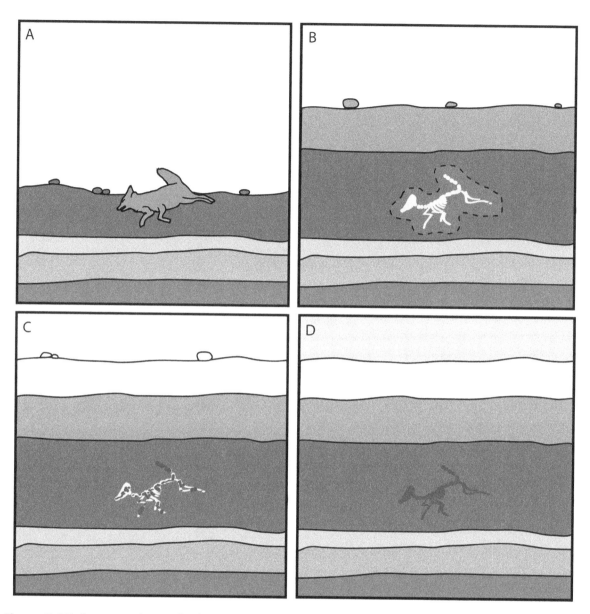

Figure 2.27: Sequence of events leading to the burial and fossilization of an animal. A) An animal has died on the Earth's surface; B) The skeleton has been buried by younger layers; C) Bone tissue begins to fossilize due to permineralization; D) Bones have now become completely replaced by new minerals.

typically relatively young, as most of the ice we see today had formed during the most recent peak in glaciation during the later part of the Pleistocene epoch.

It would be a little far-fetched to go to your local fast food chain and order a McMastodon or McMammoth, but, believe it or not, people today really have eaten the meat from these frozen ancestors of elephants. Of course, these people tend to be researchers or locals who come across the frozen carcasses and eat the meat either as a dare or out of pure necessity. The meat, which can still retain a pink, fleshy color, is described as smelling horrible and tasting just as bad.

TRAPPED IN AMBER

Amber is fossilized tree resin, which is a type of liquid, similar to sap, released by trees to cover and protect any injuries to trunks and branches. Amber shows up in the fossil record as nodules of hardened resin; not only is the amber itself

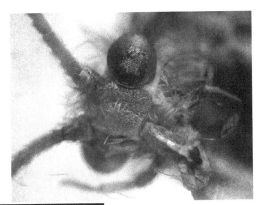

Figure 2.28: Top left: a preserved baby mammoth. Top right, a caddisfly trapped in amber. Bottom: Skeleton of a Pleistocene cave bear in a Romanian cavern. This species of bear used caves for shelter.

a fossil, but some of it contains other fossils trapped inside, such as insects (Figure 2.28B), arachnids, spider webs, plant matter, pollen, feathers, fur, and in rare cases, small vertebrates.

Amber preserves whatever it encases very well, but that doesn't mean we can extract dinosaur DNA from a Mesozoic mosquito like in *Jurassic Park* (1993). DNA breaks down as a result of water and microbes, two things very common in the ground where most fossils are buried. Even DNA trapped in amber, and tar, for that matter, is subject to degradation by enzymes already present amongst the nucleotides. Recent research has indicated that DNA has a half-life of around 521 years, precluding dinosaurs from any potential resurrection.

CAVES

Caves provide shelter for a plethora of organisms, some of which die and remain undisturbed for long periods of time (Figure 2.28C). Although most caverns and caves form as a result of water-related weathering, many can become dry, which allows for excellent preservation. In Arizona's Grand Canyon Caverns, for instance, a fifteen-foot, two-thousand-pound giant ground sloth had become permanently stuck in the cave over eleven thousand years ago. Above the skeletal remains of the sloth is a cave wall bearing numerous scratch marks left in a vain attempt to escape.

In limestone caverns (considered in this context to be a form of cave), remains of plants, insects, and animals may become encrusted with calcite and become incorporated into solid rock rather quickly. Young Earth theorists and other opponents cite this phenomenon as proof that fossils don't require long periods of time to form. It should be noted, however, that not all forms of fossilization occur in such contracted time frames, and some fossils can take millions of years to form. For example, Pliocene beach deposits are exposed along the shoreline cliffs at Capitola Beach, California. Sandstone over three million years old at this location contains shells of clams and gastropods that can be dug out from the sand and cleaned off as if

Figure 2.29: Sierran granite "decomposing", a chemical weathering process that converts feldspar into clay, and oxidizes the mafic (with an F) minerals. Note that the majority of remaining minerals are quartz, which is particular stable.

they were just plucked off the beach. The only sign that they had been around for millions of years, besides their inclusion in rocks whose age was established using a variety of relative and absolute dating methods, is a slight amount of recrystallization in the shells (Figure 2.30).

TRACE FOSSILS

Some fossils aren't truly remains of an organism itself, but are rather traces of an organism's existence (Figure 2.31). Most organisms alter their surroundings in one way or another as a result of **biological activity**. For instance, the footprints left behind by an organism can be preserved in lithified mud, silt, and sand. Tunnels left behind by burrowing organisms can also be preserved if the burrows are filled in with new sediments before they have a chance to collapse. **Coprolites**, which are fossil feces, can be fossilized, especially in dry environments such as deserts and some caves.

DATING METHODS

There are two ways to understand the timing of past geological events, such as volcanic eruptions, the formation of rocks, mass extinctions, and tectonic deformation. **Relative dating** gives us clues about the order in which past geological events occurred, but it doesn't tell us how long ago they took place. **Absolute dating**, on the other hand, provides us with a true age of a rock or geological event.

RELATIVE DATING

To interpret the geological history of the crust, geologists employ the laws set forth by Nicolas Steno and Charles Lyell. This set of rules is used while observing and interpreting rocks in the field. These laws have been integral in the formation of the geological time scale; the eras and periods you are familiar with, such as the Paleozoic or Jurassic, are based on periods of Earth's history from which particular fossils originated. By establishing the chronological order different fossils were formed in, we can observe the descent of life with modification through time, and therefore the theory of evolution depends on knowing the order in which past organisms had appeared.

Figure 2.30: This 3-million-year-old clam fossil was removed from a sandstone that still hasn't quite lithified yet.

LAW OF INCLUSIONS

Any rock that's embedded in another rock is older than the rock it is embedded in (Figure 2.31). That may sound confusing, so think of it this way: The raisins in a loaf of raisin bread are older than the loaf of bread they are in. Otherwise, the alternative scenario is that grapes spontaneously sprouted in the loaf and subsequently shriveled up.

The left-hand rock in Figure 2.32A is a chunk of basalt with inclusions of mantle peridotite embedded in it. The peridotite was already a solid rock when the basalt surrounding it was liquid, and it was plucked from the upper mantle by a rising plume of magma. The magma erupted as basaltic lava once it reached the surface of the Earth. There, the basalt surrounding the peridotite finally cooled and froze solid. The peridotite is older than the basalt.

On the right is conglomerate, a clastic sedimentary rock composed of cemented grains of silt, sand, pebbles and cobbles. It likely formed from sediments deposited along a river or beach; places where rocks are rounded by flowing water. Every single grain of sediment in this rock was weathered from distant outcrops at some point prior to the rock's formation. In that sense, every grain of sediment is thus an inclusion in the conglomerate and is older than the conglomerate itself. This principle applies to all clastic sedimentary rocks.

LAW OF CROSS-CUTTING RELATIONSHIPS

Things that cut across rocks, such as faults, folds, fractures, igneous intrusions, and zones of metamorphism, are younger than the rocks they cut across (Figure 2.33). These features can only form within existing crustal rocks.

The notion of faults being younger than the rocks they cut across is not technically true. For instance, Figure 2.33 is a photo of a sidewalk offset by the Calaveras Fault in Hollister, CA. The fault is much older than the sidewalk, which is about sixty years old. The most recent movements of the Calaveras Fault, however, have occurred after the sidewalk was paved. In other words, the crack cross-cuts the sidewalk and is therefore the younger of the two.

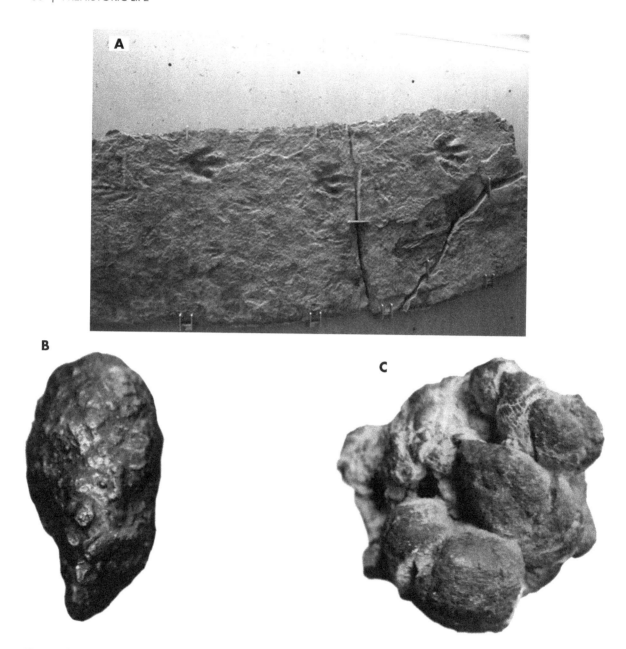

Figure 2.31: Trace fossils; A) dinosaur footprints, exhibit in Snee Hall, Cornell University; B and C) coprolites are fossil feces.

LAW OF SUPERPOSITION

A stack of rock layers, such as sedimentary beds or lava flows, will be youngest at the top and oldest at the bottom. For instance, valley floors collect sediments from the surrounding mountains. The ground surface in a valley makes up the of a large stack of sedimentary beds that are progressively older the deeper down you go. Each layer of sediment, even those now very deeply buried, was at one time forming at the surface of the Earth.

LAW OF ORIGINAL HORIZONTALITY

When layers of rock are tilted or folded, we can assume the layers were not initially deposited that way and that some tectonic disturbance reoriented the beds after they had formed. Conversely, beds of rock that are horizontal

Figure 2.32: A) Inclusions of mantle peridotite in basalt; B) conglomerate composed of sedimentary inclusions.

are assumed to be in their original orientation. This is a general assumption and doesn't apply to many examples of sedimentary beds and lava flows that did, in fact, form on a sloping surface of the crust. The bottoms of bodies of water are generally the best places to deposit horizontal layers. For example, the beds of rock seen in the walls of the Grand Canyon (Figure 2.34) were deposited mostly underwater during several marine transgressions (episodes of sea level rise that inundate the land).

LAW OF LATERAL CONTINUITY

No sedimentary or volcanic rock layer extends indefinitely; all layers have edges to them somewhere. This is because there has never been a layer of rock that blanketed the entire Earth. For instance, layers deposited at the bottom of a lake are only going to be as big as the lake ever was.

Figure 2.33: This sidewalk in Hollister, CA, has been cross-cut by recent movements on the Calaveras Fault.

Figure 2.34: Both lateral continuity and original horizontality is demonstrated in this view of the Grand Canyon, AZ. The light-colored layer of limestone the tourists are standing on is also visible towards the tops of the buttes in the distance.

It is very important, however, to be able to **correlate** sedimentary layers, which is the practice of identifying rocks of similar age, even when these rocks are separated by long distances. The law of lateral continuity is more accurate over shorter, more limited distances, such as those between sets of layers in adjacent canyons or further along the same cliff exposure.

LAW OF UNCONFORMITIES

Unconformities (Figure 2.35A) are boundaries between rocks of discontinuous age and represent a span of time not recorded in the layers. They are analogous to missing chapters in a book; you might not realize any pages are missing until you read it. Similarly, a stack of sedimentary beds may just sit there, for long periods of time, and not collect any sediments. You may not recognize this hiatus in deposition until you take a closer look at the rocks and see fossils of a twelve-thousand-year-old mastodon directly above a layer containing dinosaur bones. Where did the intervening seventy million years go? There are two reasons why this may occur:

1. There were beds from the seventy-million-year interval that had formed atop the dinosaur bones but they were subsequently stripped away by erosion prior to the deposition of the mastodon fossils.
2. No beds ever formed during that period of time.

There are three general types of unconformities (see Figure 2.35B). A **disconformity** is an unconformity between sets of layers that are parallel to each other. When the layers are not parallel across the boundary, it is called an **angular unconformity**. Finally, when sedimentary rocks are juxtaposed against igneous and/or metamorphic rocks, it is called a **nonconformity**.

ABSOLUTE DATING

For the latter part of the nineteenth century, the geologic time scale did not attempt to suggest with any accuracy the lengths of time each period was or how long ago they occurred. It wasn't until the early part of the twentieth century that geologists came up with techniques allowing them to take a single crystal from a rock and determine its true age in actual years.

Absolute dating refers to any technique that gives us an indication of how old something is. Many of these methods rely on the measurement of radioactive elements in a sample of rock or organic matter.

In 1896, in his lab in Paris, a French physicist and chemist named Henri Becquerel discovered that uranium collected from rocks emits its own energy. He had been researching the ability of crystals to absorb solar radiation and had previously thought that energy he observed radiating from uranium had only been stored there temporarily after the sample had sat in the sun.

The energy was detected by placing photographic plates near samples of uranium he had derived from rocks. The uranium created a light, invisible to the human eye, that could only be seen as distinct shapes on a photograph developed from the plates (Figure 2.36). This observation had been made about fifty years earlier by photographers, who had discovered that uranium salts were useful for creating a decorative darkened border around photographs, but Becquerel was the first to publish these observations in a scientific context.

In 1902, New Zealand-based physicist Ernest Rutherford (1871–1937) (Figure 2.37) discovered that rocks containing uranium and other radioactive elements seemed to always contain lead as well. He realized five years later that not only was the uranium turning into lead, but it was doing so at a measurable rate. This observation and the assumption that the lead in the samples was produced at the expense of uranium suggested that the actual ages of rocks could be determined. The method involves measuring the ratio of uranium to lead that they contain. (See Chapter 5 for more details). The accuracy of this technique is primarily based on how well we can measure the amounts of various elements in a sample.

By 1907, Rutherford had dated enough rocks to infer the age of the Earth to be 2.2 billion years old. The establishment of this age, although less than half of the age we believe the planet is today, was a huge step forward in understanding the past, and the techniques developed by Rutherford have been honed over the last century and are still in use today.

THE GEOLOGIC TIMESCALE

Since the late 1700s, naturalists had been describing sets of rocks in various locations, mostly throughout Europe. These early groupings of rocks were called **systems** and were based primarily on rock type; most descriptions didn't focus much on any fossils that may have been present. The names of the systems typically came from local cities, regions, and sometimes people. The systems determined back then were used to describe rocks at a particular place, and the names and descriptions used were unique to that location and did not apply anywhere else.

For well over one hundred years, naturalists were naming numerous sets of rocks but had no clear idea how they related to one another. In fact, some systems overlapped. A system of rocks in northern Wales was

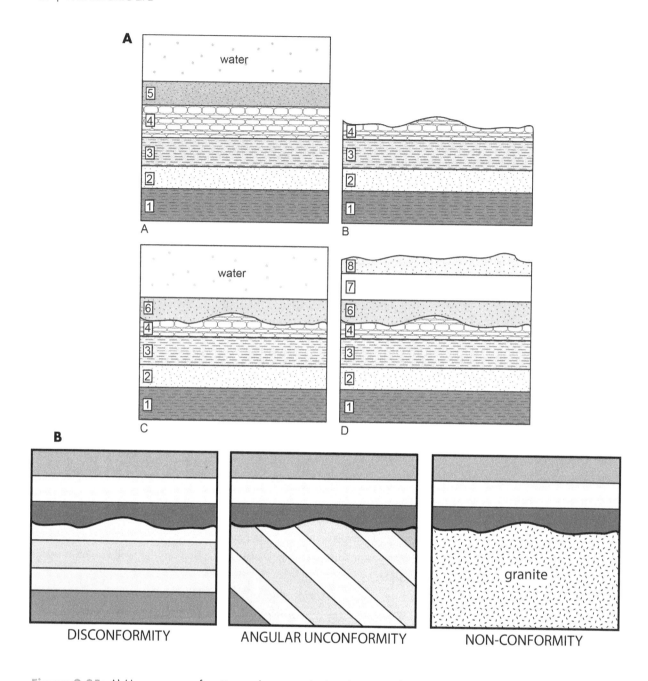

Figure 2.35: A) How an unconformity can be created; a) sedimentary layers accumulate for a period of time; b) for millions of years, erosion strips away some of the layers; c) and d) sediments begin to accumulate once again. B) Illustrations of common types of unconformities.

described in the 1830s by Adam Sedgwick, who named the rocks **Cambrian**. Concurrently, Sir Roderick Murchison described a set of rocks to the southeast of the Cambrian rocks and called them **Silurian**. Having foresight, Murchison took the extra time to describe the fossils he found in the rocks in addition to the lithology. Both men published descriptions of their systems in 1835 and were annoyed to see that the top of the Cambrian system correlated with the bottom of the Silurian system; this might not have been realized had Murchison ignored the fossils in his rocks. A source of contention, this overlap represented rocks that both Sedgwick and Murchison claimed to be part of their systems. They fought over this for another forty

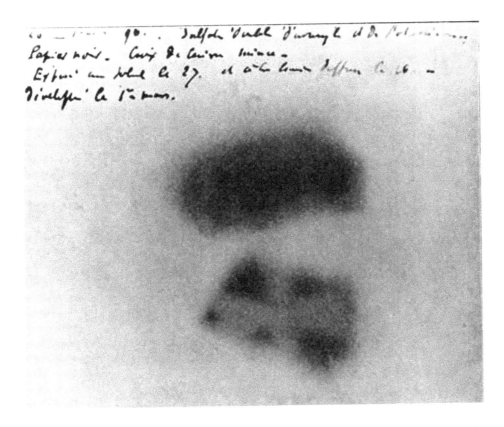

Figure 2.36: Photograph that prompted Henri Becquerel to realize that uranium was emitting its own radiation.

years until another geologist, Charles Lapworth, suggested the overlapped zone should be its own system called the **Ordovician.**

Subsequent recognition of new systems was aided by the study of the fossils they contained. It was realized early on that one could not time-correlate a set of rocks between two locations without looking at the fossils they contain. The early geologic timescales were created by the stitching together of these different systems into chronological order based on their relative ages and confirmed by the fossils they contain. Now, one could recognize Cambrian rocks in many locations by the fossils they contain, not by the type of rocks they are. The fossils allow us to determine **time-equivalence** between sets of rocks around the world; rock type alone does not.

Prior to the establishment of absolute dating methods, the geologic timescale didn't indicate true ages of rocks. For example, scientists had been describing rocks from the Jurassic period since 1795, but they didn't know when the Jurassic had actually taken place until the utilization of absolute dating methods (see Figure 2.38) in the early twentieth century. Early iterations of the scale organized the rocks based on their relative positions and, not long after, by the fossils they contained. Today, fossils are the primary line of evidence used to date the rocks that contain them.

Figure 2.37: Ernest Rutherford developed the first absolute dating methods utilizing isotope concentrations from mineral samples.

HOW TO CALCULATE ABSOLUTE AGE

Sometimes when crystals form, such as those in igneous rocks, unstable isotopes may be absorbed into the crystal structure. These are called PARENT ISOTOPES.

Over time, these unstable isotopes radioacively decay into a more stable atom, called a DAUGHTER ISOTOPE.

Each type of isotope decays at a particular rate, referred to as a HALF-LIFE. Half of the parent isotopes in a crystal will decay during each half-life. To the right is a table that shows how much of the parent and daughter isotopes there will be in a sample after each half-life. This applies to any isotope that exhibits half-life behavior.

RELATIVE AMOUNTS OF PARENT AND DAUGHTER ISOTOPES AFTER THE FIRST FIVE HALF-LIVES

Number of Half-Lives Elapsed	Amount of Parent isotope	Amount of Daughter isotope
0	100%	0%
1	50%	50%
2	25%	75%
3	12.5%	87.5%
4	6.25%	93.75%
5	3.125%	96.875%

We can measure the relative amounts of parent and daughter isotopes in a sample to determine how many half-lives have elapsed since the crystal formed. We already know how long the half-life is, so we canthen calculate how old the sample is.

In most introductory Earth Science classes, you will be asked to determine the age of a sample using this method. For instance, if you know the ratio of parent to daughter isotopes, and the length of the isotope's half-life, you can determine age. Below is an example:

A sample contains 12.5% of it's original parent isotope. The isotope being measured has a half-life of 2,000 years. How old is the sample?

Step 1: Refer to the table and determine how many half-lives have gone by. In this example, when there is 12.5% of the parent isotope left in the sample, it means that 3 half-lives have gone by.

Step 2: Multiply the number of half-lives by the length of the half-life. In this case, we multiply 3 half-lives by 2,000 years for each half-life. 3 x 2,000 years = 6,000 years. The sample is therefore 6,000 years old.

Figure 2.38: A) Table of Parents and Daughters after X numbers of half-lives have elapsed.; B) Table of well-used isotopic decay pairs; C) How to calculate the age of a sample based on this data.

It was only after these systems of rock were organized chronologically and their fossils examined that scientists began to see progressive changes in the types of fossils through time, as well as episodes of sudden diversification and, conversely, mass extinctions. So began the timescale and the framework in which the study of evolution could take place.

Geological time is broadly separated into two **eons**. The first eon is known as the **Precambrian Eon**, beginning when Earth began and lasting all the way until 542 million years ago. At this time, when the **Phanerozoic** Eon began, life on this planet was in the early stages of evolving into many of the major phyla we still have today. The meaning of *phanerozoic* loosely translates to "visible life." Both eons are further segmented into **eras**. The Precambrian Eon consists of the **Hadean, Archean,** and **Proterozoic Eras**. The Phanerozoic Eon is divided into three eras: **Paleozoic** (early life), **Mesozoic** (middle life), and **Cenozoic** (recent life).

Each era is divided into distinct **periods**. For instance, the Paleozoic is made up of several periods: **Cambrian, Ordovician, Silurian, Devonian, Carboniferous** (which in the U.S. is further divided into the **Mississippian** and **Pennsylvanian** periods), and the **Permian**. The Mesozoic is made up of the **Triassic, Jurassic,** and **Cretaceous** periods. Finally, the Cenozoic is divided into **Paleogene, Neogene,** and **Quaternary** periods.

Each period is divided into epochs, and each epoch is divided into **ages**. Ages are very specific slices of time, and there are so many that they don't fit on a normal-sized time scale. Today, we belong to the Phanerozoic Eon, Cenozoic Era, Quaternary Period, and Holocene Epoch. As you read through this book, more details will be filled in about these divisions of time mentioned above. *The last page of this book has a timescale for your reference.*

THE ANTHROPOCENE

Millions of years in the future, scientists, whether our own descendants or not, will not have a hard time identifying layers of rock deposited during the time when humans reached their apex. The record of our existence will be a set of layers with a high concentration of radioactive material, a multitude of volatile chemicals, glass, concrete, metal, and other synthetic residues. Future scientists will notice that many different organisms went extinct right at this boundary.

Because we will no doubt leave our mark on this planet, many scientists agree that we have, in fact, entered a new epoch, called the **Anthropocene**, named after us. It has yet to become officially adopted, however, mainly because there is debate about exactly when this new epoch began.

3

Figure 3.1: A statue of Louis Agassiz took a nose dive into the concrete during the 1906 earthquake, Stanford University, Palo Alto, CA.

STRUCTURE OF THE EARTH

THE LITHOSPHERE AND ASTHENOSPHERE

The Earth's crust varies in thickness between 5 and 125 kilometers and is distinguished from the mantle based on its composition. One critical difference is the amount of **ferromagnesian** minerals present in each of these layers. These minerals contain iron and magnesium as part of their chemical makeup, which makes them relatively dense and dark in color and gives them higher melting temperatures than many other minerals. In granite, for example, the black-colored minerals (e.g., biotite and hornblende) are examples of ferromagnesian minerals.

When the Earth was still a molten ball of magma just after it reformed (we believe it had formed once before, then was impacted by a Mars-sized planet and had to start over), it didn't have the distinct concentric layering we see today (Figure 3.2). While still molten, the mixture of lightweight gases, heavy metals, and everything in between began to **differentiate**, a process that, in this case, involves the separation of substances with different densities as a result of gravity. Many of the heavy elements, such as iron and nickel, sank to the middle of the molten Earth and formed the core, while most of the ferromagnesian minerals, not as dense as the pure metals, formed the mantle, atop the core. Lighter-weight minerals rose to the top of the mantle and cooled, forming a primordial crust.

In the Earth, both temperature and pressure increase with depth. The heat from the Earth's initial formation has likely dissipated already, yet the Earth is still very much hot inside. This is because our planet is warmed by the energy escaping from radioactive isotopes in the core. The nuclear decay of these unstable atoms is exothermic, releasing tremendous amounts of heat that radiate from the core and eventually escape the crust as geothermal energy.

The **geothermal gradient** describes the increase in temperature with depth. Away from plate boundaries, the crust increases by about 1°F for every seventy feet of depth. Some of the deepest mines on our planet reach nearly 2.5 miles below the surface; air temperatures in these mines can reach up to 130°F, making heat exhaustion a real threat to miners in the deeper workings.

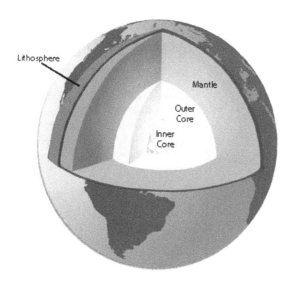

Figure 3.2 Major divisions of the Earth's interior.

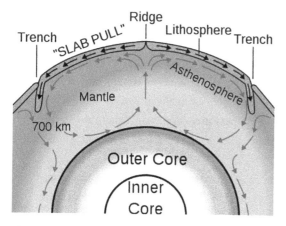

Figure 3.3: Mantle convection is a driving force behind the movement of the plates.

Depending on location, the depth to which rocks are hot enough to melt varies between 50 and 125 miles, in the upper mantle. Farther down, at about 400–500 miles deep, the pressure is so high that rocks cannot melt, since they need to expand into a liquid state. Although temperatures keep rising below this depth, most of the mantle remains solid due to high amounts of pressure.

This interaction of heat and pressure allows a narrow zone within the upper mantle to be hot enough to melt, yet not under enough pressure to prevent it from melting. This zone of partially melted rocks is called the **asthenosphere**. Unlike ice, a pure substance, most rocks are composed of a variety of minerals, each with a different melting temperature. The heat and pressure conditions in the asthenosphere allow just enough melting of felsic minerals to occur so that it is soft enough to flow and behave like a liquid.

Above the asthenosphere is the **lithosphere**, which is composed of the crust and the uppermost solid mantle. These two layers are stuck together, forming a rigid slab of rock that more or less floats and glides atop the asthenosphere (Figure 3.3). The lithosphere is broken into several fragments called **tectonic plates** that crash together, pull apart, and move alongside each other.

One of the first maps (Figure 3.4) to show the coastlines on either side of the Atlantic was produced in 1529 by Diogo Ribeiro, a Portuguese cartographer hired by the Spanish. One thing that stood out in his map was the similarity between the shape of the coastlines on

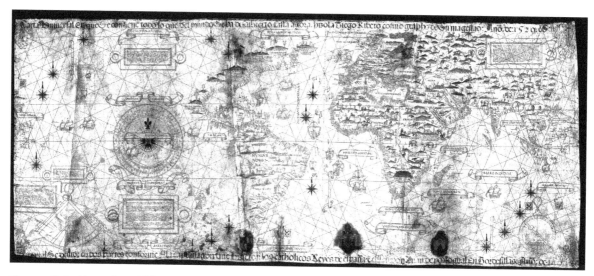

Figure 3.4: Map of world from 1529 by Diogo Ribeiro.

either side of the Atlantic. This similarity is fairly obvious and implies either coincidence or that the continents were at one time together and either got flooded or broke apart along the Atlantic coastline. At the time, the latter explanation was very hard to justify; it seemed ridiculous to most people that there could be some force moving huge slabs of the Earth's surface around, overcoming great amounts of friction to do so. The only explanation that made any sense to people was that the Atlantic formed as an effect of the great biblical flood.

The theory of continental drift, although made famous by Alfred Wegener in the early twentieth century, was first postulated in 1596 by Abraham Ortelius, a Flemish geographer and cartographer, in his book, *Thesaurus Geographicus*.

Over the next few centuries, discoveries were made that lent more credence to the idea of moving continents. In 1858, a French geographer named Antonio Snider-Pellegrini published a book titled *The Creation and Its Mysteries Unveiled* in which he proposed that the continents on either side of the Atlantic were once connected in the Pennsylvanian period. One line of evidence he used was the presence of identical plant fossils in Europe and the U.S., implying the two areas where the fossils are found today were once together, or at least not separated by an ocean.

By 1885, Eduard Suess, an Australian geologist, had published research on one of these plants, a ubiquitous fossil fern *Glossopteris* from locations found throughout the southern hemisphere. He concluded that they represent what was once a contiguous population of plants that stretched across modern-day South America, Australia, Africa, India, and Antarctica. Suess proposed that the Atlantic and other southern oceans had flooded the supercontinent, leaving isolated bodies of land, not that the lands had physically moved apart.

In 1910, F.B. Taylor, an American physicist, attributed the formation of mountain belts to giant crustal wrinkles formed when the crust gets shortened. This theory is fairly close to what we understand today as a reason for mountain building. He and another colleague, H. Baker, also noted that some mountain ranges, such as the Appalachians, are old and appear to represent fragments of an ancient mountain range now broken apart and separated by widening seas.

Throughout the long period of time between the mid-1500s and the early 1960s, there were many theories published about giant floods and drifting continents, but none of them were able to explain in a scientific sense *why* or *how* this was occurring. Furthermore, some of the explanations were outlandish, such as one touted by H. Baker in the early 1900s. His explanation was that Venus had passed very close to Earth and its gravity caused a large patch of the crust to peel off, float up into space, and form the moon.

Despite multiple lines of evidence that indicate the continents move, the inability of scientists to explain the mechanisms allowing drift to occur prevented this theory from becoming widely accepted. In fact, many early scientists in this field didn't receive proper recognition for their contributions until long after they were dead.

One of the scientists most famous for his continental drift hypothesis was Alfred Wegener, a German geophysicist, astronomer, and meteorologist (Figure 3.5). We recognize the importance of his (and his predecessor's) research today, but while he was still alive (he froze to death during an expedition in Greenland), the scientific community scoffed at his ideas. He was the first to publish a *comprehensive* report describing the multiple lines evidence of continental drift collected and published by earlier scientists such as Suess and Taylor, as well as his own contributions. It was through his research that we were formally introduced to the supercontinent of **Pangaea**.

Figure 3.5: Alfred Wegener (1880–1930).

ALFRED WEGENER AND CONTINENTAL DRIFT HYPOTHESIS

Wegener was aware that many of his colleagues, some of them world-renowned in the fields of physics, geology, and meteorology, thought he was wasting his time. Nevertheless, he continued with his research despite discouragement from the "great thinkers" of the day. Herein is one of the issues with science: Scientists can be very skeptical of research that attempts to upset the status quo.

Many of the greatest discoveries in science were made by people who devoted their lives to pursue theories that most other scientists refused to consider because they were biased by scientific dogma, theories so well-established that they are viewed as fact. The idea of a static, stationary crust had been so ingrained in the minds of even the greatest thinkers that they didn't dare to contradict it, especially if nobody could ever explain why or how continental drift occurred. Wegener could not explain it either, but his collection of evidence supporting continental drift, published in 1915, later provided the basis for **plate tectonic theory**, which began to form thirty years after Wegener's unfortunate death.

Wegener sited several convincing lines of evidence that support the **continental drift hypothesis**:

1. *The puzzle-like fit of the continents*, especially on either side of the Atlantic, had been noticed early on and was the primary reason why so many scientists and cartographers since the 1520s had inferred the possibility of crustal movement. Wegener coined the term "Pangaea" (meaning "all Earth") as a name for the supercontinent implied by these lines of evidence.

2. *Identical fossils of plants and animals* found on either side of the Atlantic indicate the continents they are found on today were once together (Figure 3.6A). Examples include the fern-like *Glossopteris* mentioned above and the Mesosaurus, an extinct freshwater reptile from the Early Permian that predates dinosaurs and is found in rocks of both South America and Africa. Fossils of the Brachiosaurus (more modern name for Brontosaurus) can be found in both Colorado and Tanzania.

3. *Living organisms on either side of the Atlantic today appear to have common ancestors*. Examples include marsupials, which are found in North America, South America, and Australia, and earthworms of the family Megascolecidae, which are found across South America, Madagascar, Africa, Australia, and India. When Pangaea is reconstructed, groups of fossils of various organisms form cohesive patches of land representing the ancient habitats of these creatures as they were in the Mesozoic.

4. *Geologic structures such as mountain ranges, thick sequences of sedimentary beds, and metamorphic belts are seen in places to truncate at high angles against coastlines,* as if they were part of some contiguous structure that diverged along with the continents they formed on. The Appalachian Mountains are the remnants of a much larger mountain range, similar in size to the Himalayas, that had formed during the collision of what is now the eastern U.S. and Africa during the Carboniferous. Mountain belts of equal age, level of metamorphism, and sequences of rocks are also found in the Cevennes range of France and the Carpathian range of Eastern Europe today. The presence of these ancient, fragmented ranges also implies that Pangaea was constructed by the collision of older, pre-existing continents.

5. *Evidence of past glaciers, tropics, and deserts are found at latitudes today where we wouldn't expect them to form.* The Earth currently has two continental glaciers: one on Greenland, and one on Antarctica. These are located near the poles and form when snow accumulates into thick deposits and compacts to form glacial ice. Large continental glaciers such as these are ductile (capable of flow) and slowly spread outward in all directions under the influence of gravity. This creates a recognizable large-scale pattern of radiating scratch marks on the crustal rocks beneath the ice, as well as a concentric distribution of glacial sediments (Figures 3.6B and 3.7). We see evidence today of an ancient ice cap in places such as India, Antarctica, South America, Africa, and Australia. When these continents are fitted back together, these glaciated areas reform into a cohesive patch of crust located at the South Pole. We also observe desert sandstones, fossils of tropical plants, and coal deposits in locations too cold for them to form in today, such as Antarctica and northern Europe.

Alfred Wegener presented these and other observations as evidence for continental drift, but he failed to convince many people of his hypothesis due to his inability to explain the driving forces behind the movement. Two major questions were difficult to answer at the time:

1. How is it possible to overcome the huge amounts of friction expected when rocks slide on top of other rocks? and

2. What is causing the continents to move?

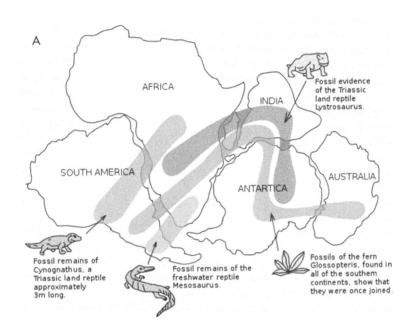

Figure 3.6: Two lines of evidence used by Wegener to support his Continental Drift Hypothesis: A) Similar fossils on separate continents; B) Glacial striations in Brazil from when South America was still connected to Africa near the South Pole

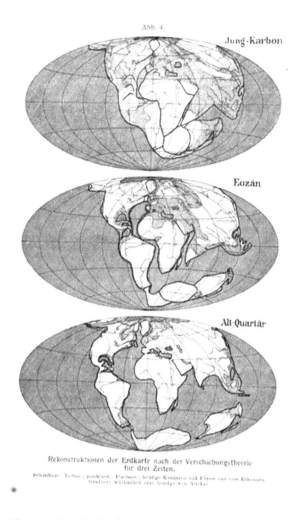

Figure 3.7: Early Illustration of Pangaea by Alfred Wegener.

The answer to both of these questions came once we had a better understanding of the interior of the Earth and the shape of the ocean floor. A new surge of evidence collected during the mid-twentieth century shed much more light on possible mechanisms behind continental drift.

PLATE TECTONICS THEORY

It is understandable that Wegener and his predecessors couldn't adequately explain the mechanisms responsible for continental drift. This is due in part to the lack of knowledge about the ocean floors, which constitute nearly 70 percent of the Earth's surface. Furthermore, not much was known about the interior of the Earth, which according to this theory, must contain some interface upon which the continents can slide. It would be like trying to understand internal human physiology by only looking at a person's skin.

Beginning in the mid-twentieth century, two major discoveries shed more light on the structure of the Earth. The first discovery came about during a period of intensive sonar studies of the ocean basins. A collaborative effort of military ships and research vessels resulted in high-resolution maps of the ocean floor. To our surprise, the floors of the oceans weren't simple, flat expanses of crust, but rather a terrain dotted with undersea volcanoes, long, continuous volcanic ridges, and deep trenches. Even more surprising was that the ridge mapped in the mid-Atlantic appears to have the same shape as the coastlines on either side and is located almost exactly in the middle of the basin.

The second discovery was that we could use seismic waves to resolve density differences in the Earth's interior. This ability was discovered just after the largest earthquake in recorded history (M8.5) struck Chile in 1960 (see Figure 3.12). It was observed that seismic waves passing through the interior of the Earth slowed down in places, suggesting the waves were encountering low-density, partially melted rocks. In particular, at a depth of about 180–220 kilometers below the surface, seismic waves encountered soft, partially melted rocks and slowed down, causing them to arrive "late" at distant seismometers. Although some scientists had already began to speculate the existence of the asthenosphere as early as the mid-1920s, it wasn't until 1960 that we were able to show with concrete data the existence of a low-velocity zone (LVZ) in the upper mantle. It was previously assumed that the continents would have to slide on top of solid rocks in the mantle to be able to move, and the friction involved would simply be too great to allow this to happen. With this new evidence-backed knowledge of the asthenosphere, we were able to conclude that there wouldn't be as much friction involved if the lithosphere was sliding on a hot, slippery layer of mantle below the crust. This didn't quite explain, however, what's causing the movement to begin with.

Once the semi-liquid nature of the asthenosphere was discovered, a logical explanation for the movement of the plates arose. Because the mantle below the asthenosphere is solid, heat from the Earth's core **conducts** outwards until it reaches the upper mantle. Being more like a liquid, the asthenosphere **convects** the heat upward by *flowing*. Convection cells allow denser, cooler asthenosphere at the top to sink and warmer, more buoyant asthenosphere at the bottom to rise, completing a gyre of rotation that helps the heat efficiently escape the mantle.

THE TWO TYPES OF LITHOSPHERE

Rocks are made of minerals, and minerals have a variety of compositions and characteristics (Figure 3.8). Some of these minerals contain within their chemical formula iron and magnesium, which makes them dense, dark in color, and have high melting/freezing temperatures. These minerals are called **mafic**, or **ferromagnesian** minerals.

Mafic minerals make up most of the Earth's mantle and are also present in the crust in varying amounts. These minerals combine to form mafic rocks, which can be described as rocks that contain relatively large amounts of mafic minerals.

Rocks that form the oceanic crust are mafic, which makes the oceanic crust dense and dark in color. This also makes the oceanic crust relatively less buoyant than continental rocks. Oceanic crust is also the only type of crust produced at **divergent boundaries**, and it is rather thin (four to seven kilometers).

Despite the name, oceans have nothing to do with the formation of oceanic crust. Because the crust forming at divergent boundaries is thin, dense, and mafic, it sinks down further into the asthenosphere and forms deep basins that cover over 70 percent of the earth's surface. The oceans simply flowed by gravity into these basins.

Non-ferromagnesian (aka **felsic**) minerals lack iron and magnesium and are mostly made of silicon, oxygen, and other lighter-weight elements. These minerals are found in much higher concentrations in **continental crust**. Continental crust, much thicker than oceanic crust is produced at **convergent plate boundaries** and forms thick slabs of lightweight, relatively buoyant rock.

MAFIC

FELSIC

Figure 3.8: Examples of the mafic mineral biotite (left) and the felsic mineral quartz (right).

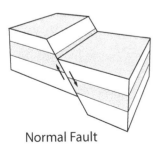
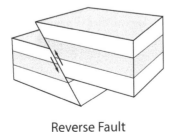
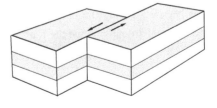

Normal Fault

Reverse Fault

Strike-Slip Fault

Figure 3.9: Three block diagrams illustrating different fault types.

Since divergent boundaries only involve the production of oceanic crust, the interactions of oceanic crust and continental crust are mostly limited to convergent boundaries (Figure 3.10).

DIVERGENT BOUNDARIES

Divergent boundaries are where new oceanic crust is produced, and the plates on either side move away from each other like two opposing conveyor belts. This process is called **seafloor spreading** and is the reason why continents split apart.

Because intermittent sheets of molten rock rise and fill in the spreading gap between the plates, minor amounts of volcanism are associated with these boundaries. The movement of magma as well as the differential rates of spreading along the boundary results in minor to moderate earthquakes as well. Spreading rates are typically slow (only a few mm/year).

The constant generation of new crust along divergent boundaries creates ocean basins whose rocks are youngest at the site of spreading and are progressively older away from the ridge.

Eventually, all divergent boundaries become surrounded by the oceanic lithosphere they create, but they may initiate within a continent at first. When this happens, a **rift valley** is formed; examples include the East African Rift Zone, the Red Sea, and the Gulf of Aden. Rift zones begin as linear valleys of sunken, thinning, sinking continental crust that typically contain lakes and are dotted with volcanoes. As divergence continues, thin oceanic crust forms along the axis of the valley, and when the valley sinks below sea level, it becomes flooded with seawater and a new ocean begins to form. This is how the Atlantic Ocean initiated about 180 million years ago; Pangaea "unzipped" from north to south, finally tearing completely apart after about 100 million years of divergence. The age of the Atlantic sea floor is oldest in the North, indicating that's where spreading first began in earnest.

Once a divergent boundary is surrounded by oceanic crust, it generally forms an undersea ridge cut by perpendicular transform faults, giving it a zigzag pattern. Differential rates of spreading and the curvature of the Earth prevent the ridge from spreading evenly. It may sound counter-intuitive that a plate boundary defined by rifting would be marked by a ridge at all, but there are three reasons why it stands out above the surrounding seafloor.

1. Nearest the ridge, the crust is thermally expanded because it is still hot. This makes it thicker than the surrounding rocks. Eventually, the rocks cool as they move away from the ridge and shrink as a result.

2. Due to the thermal expansion, the crust nearest the ridge is the least dense and therefore more buoyant, allowing it to "float" higher on the asthenosphere than the cooler crust farther away.

3. Magma accumulates beneath the ridge, pushing the crust upwards.

The oldest rocks composing oceanic crust are about 180 million years old, indicating that some other mechanism has been destroying the older crust. It is also logical that if new crust is being produced, it must also be destroyed in equal amounts, since the Earth can neither inflate nor deflate.

CONVERGENT BOUNDARIES

There are two types of crust (oceanic and continental); therefore, there are three different combinations of crust that can crash together. Two of these combinations involve at least one plate of oceanic crust, and the third involves only continental rocks. When oceanic crust is involved, **subduction** occurs (Figure 3.10).

Subduction zones are convergent boundaries where oceanic crust gets shoved downwards into the asthenosphere as the other plate plows into it. These types of boundaries produce the largest earthquakes and are associated with explosive volcanoes as well as giant tsunamis. Much of the Pacific Rim is lined with subduction zones and is thus called the "Ring of Fire."

The down-going oceanic plate tends to dive at a high angle when the subducting crust is old and dense; young ocean crust tends to subduct at relatively low angles, creating volcanoes and mountains much farther inland on the overriding plate. Friction between the top of the downgoing plate and the asthenosphere

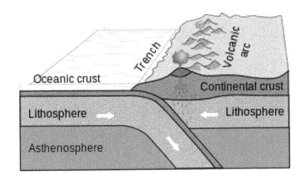

Figure 3.10: Subduction zone where two plates are colliding; the oceanic lithosphere gets pushed beneath another plate, and this results in a volcanic arc.

above it creates very large, deep earthquakes (Figure 3.11). The quakes occur on a dipping seismic surface that deepens beneath the plate above it. Between 600 and 700 kilometers deep, the seismicity ceases, indicating a level where the downgoing plate is soft and reincorporating back into the mantle. In a nutshell, the oceanic crust constitutes the frozen upper parts of large convection cells that penetrate down into the deep asthenosphere.

Conditions are right for the accumulation of juicy pockets of magma in the space between the downgoing plate and the overriding lithosphere, called the **mantle wedge**. Magma generation is aided by a depressurized zone within the mantle wedge (caused by the suction of the sinking plate) and by water, which evaporates out of the downgoing plate and aids in the loosening of superheated rock molecules into a liquid state. This magma rises and pools beneath the overriding plate, and its heat melts the crust above it, causing sheets and blobs of liquid rock to slowly work their way upward into the cooler, more solid continental crust. There, the magma accumulates in complex chambers miles below the surface and sends up incremental spurts of magma that erupt out of the Earth's crust, forming a volcanic **arc**. An arc is a curved line of volcanoes that lies parallel to the boundary. When two

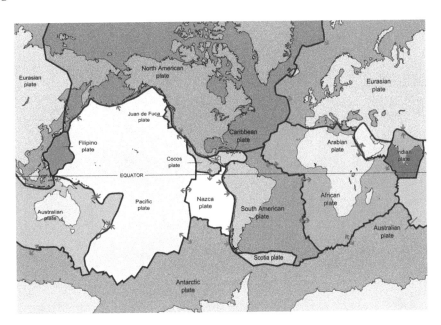

Figure 3.11: World map showing tectonic plates. The red arrows indicate the type of movement along each boundary. Note that many plates are composed of *both* continental and oceanic crust.

oceanic plates converge, the older, denser plate subducts, and an **island arc** forms in the ocean. Arcs always form on the overriding plate.

Where the oceanic plate begins to flex downward just prior to subducting, a **trench** is formed right along the boundary of the two plates. The deepest trench is the Mariana Trench in the West Pacific, whose bottom is over ten kilometers below sea level. Trenches lie along all subduction zones and tend to accumulate sediments. Some of these sediments, in addition to volcanic islands scraped off the oceanic plate as it subducts, become accreted to the overlying lithosphere. Even fragments of the oceanic crust itself, called **ophiolites**, break off and accrete to the edge of the overlying plate. When the overlying plate is continental in composition, the leading edge upon which rocks accrete becomes smashed and buckled, resulting in numerous faults and folds and a series of hills. This belt of chaotic, crumpled rocks generally has two names: the **fold and thrust belt**, and the **accretionary prism**.

Forearc and *backarc basins* lie on either side of the volcanic arc. The forearc basin forms almost by default as the volcanoes rise on one side and the fold and thrust belt rises on the other.

The leading edge of the overriding plate also tends to flex downward, stuck to the subducting oceanic plate. Eventually, the friction keeping them stuck gives way and the upper plate rebounds upward, causing the seafloor to rise suddenly and a large mound of water to form and spread outward in the form of a tsunami. Two significant tsunamis have occurred recently. On December 26, 2004, an 8.2 earthquake off the coast of Sumatra caused tsunamis over thirty feet high to spread across the Indian Ocean, causing nearly a quarter of a million deaths. On March 11, 2011, an 8.0 earthquake struck off the coast of North Japan, resulting in tsunamis over 130 feet high that killed over fifteen thousand people.

COLLISIONAL CONVERGENT ZONES

When two continental plates collide, neither subducts, as both are too buoyant and thick. Instead, they crumple upwards by folding and faulting, creating thick, tall, and wide mountain ranges, such as the Himalayas (Figure 3.12), which are called **collisional convergent** boundaries. Moderately sized earthquakes are typical, as is

Figure 3.12: The Himalayas from space; an example of a collisional boundary between two continental land masses; India and Eurasia.

some volcanism, although rising magma tends to freeze solid before it makes it to the top of the extra-thick crust. Remnants of an old, extinct collisional boundary from the formation of Pangaea, such as the Appalachian Mountains and the Scottish Highlands, are present on either side of the Atlantic, as discussed earlier.

TRANSFORM PLATE BOUNDARIES

Transform plate boundaries are not caused by tension or compression, but rather shear stress. Shear stress causes plates to slide past each other, resulting in horizontal offset of the crust. The most famous transform plate boundary is the *San Andreas Fault*, which allows the Pacific Plate to

Figure 3.13: Destruction from the 1906 San Francisco Earthquake, which occurred on the San Andreas Fault, a transform plate boundary.

slide northwest relative to the *North American Plate*. This causes Los Angeles and San Francisco to get about an inch-and-a-half closer to each other every year. Moderate earthquakes occur along the length of the boundary, and hills and valleys tend to form along bends or stepovers in the associated faults.

During the several minutes that the 1906 San Francisco earthquake (Figure 3.13) lasted, some streams, fences, paths, and other features that crossed the fault showed up to twenty-eight feet of displacement! In the section of the fault between Hollister and Parkfield, California, the San Andreas cuts through a soft, slippery bedrock called *serpentinite*, and as a result moves rather smoothly with decreased friction. In Hollister, for example, streets, sidewalks, fences, and even houses straddle the Calaveras fault and are being slowly ripped apart (see Figure 2.33).

Figure 3.14 Destruction from the 1960 Valdivia Earthquake in Chile.

4

Figure 4.1: The venus flytrap, a plant that has adapted a carnivorous diet. It's a great example of traits that were shaped by natural selection over broad spans of time.

NATURAL SELECTION AND EVIDENCE OF EVOLUTION

EVOLUTIONARY THEORY

Up until around the turn of the nineteenth century, most naturalists considered the diversity and complexity of life to be the work of God. The unfortunate reality for the fields of geology and paleontology is that scientists have traditionally been urged to conform to a framework of history consistent with biblical timelines.

The practice of identifying differences and similarities between organisms predates evolutionary theory by thousands of years. Classifying organisms into different groups is itself not very threatening to the notion of creationism; however, in combination with the fossil record, it allowed us to observe patterns of organization that suggest organisms are related not just spatially, but temporally. By the mid-nineteenth century, natural explanations for the patterns seen in the fossil and living record of organisms were being published by evolutionary biologists.

Long before the fields of biology, geology, and paleontology existed, humans had already made observations of variations within species. A **species** is a population of organisms capable of interbreeding and includes members that appear superficially similar and vary only slightly. We, as a species, have been the first on this planet to domesticate animals and grow crops, a practice that recognizes the variation in other species and uses it to our advantage. **Artificial selection** is the evolution of a species due to selective breeding of organisms with traits we prefer.

Dogs, for instance, appear in numerous breeds that are, for all intents and purposes, human inventions. Tens of thousands of years ago, humans were already skilled hunters and had eradicated many of the large predator populations they competed with for food. It's likely that ancient garbage dumps were popular spots for desperate wolves to visit and interact with people. Some wolves would have been less aggressive than others and possibly comfortable enough to join the "pack" of humans rather than continue to compete for meat. It is possible wolves tagged along during hunting trips with the expectation of getting food that didn't have to be fought for, giving them and their human companions a mutual goal of acquiring prey. If this was the case, it can be inferred that humans would have noticed that some wolves had particular traits allowing them to hunt a specific prey and would have matched up particular wolves to produce more offspring with

Canis lupus familiaris

Canis lupus lupus

Figure 4.2: Ruby, a miniature Dachshund (left) and a wolf (right) have diverged in form due to selective breeding, but still share a majority of characteristics.

those traits. It wasn't long before these inbred wolves began to appear drastically different from their ancestors (Figure 4.2).

Most of the fruits and vegetables you buy at the farmer's market or grocery store are not found like that in nature, and some are completely contrived. The plant *Brassica oleracea*, for instance, is a wild cabbage from the mustard family cultivated as far back as several thousand years ago (Figure 4.3). By cross-breeding variations of this common stock, humans have developed broccoli, cauliflower, brussels sprouts, kale, and "modern" cabbage.

Thus, the observations of variation within a species and the effects of artificial selection were well known prior to the theory of evolution (Figure 4.4). Although domestication and cultivation has led to new varieties of plants and animals, these practices tend to reduce diversity as particular traits get favored and are forced upon the population as a whole. For example, the Irish Potato Famine of the mid-1840s resulted from a blight that wiped out most of the potato crops. Only a handful of varieties were being grown, none of which were resistant to this particular disease.

By the mid-1800s, scientists began to look in earnest for natural mechanisms resulting in the variety of life forms today and began to consider the role variation within a species played in the history of life. Variation implies that some members of a population will, by chance, have more success in breeding than others due to having particular traits that enable them to better survive to adulthood and/or attract a mate. Therefore, the largest strides made since the mid-1800s have been in our understanding of the interplay between organisms and their environment.

Jean-Baptiste Pierre Antoine de Monet, Chevalier de Lamarck (Figure 4.5) was a french naturalist who, over fifty years before Darwin's *Origin of Species*, had made the inference that organisms change when their environments change and become more complex as they do so. Lamarck postulated that the adaptations seen in organisms are acquired during their lifetimes and then passed to subsequent generations. Giraffes, he suggested, lengthened their necks while reaching for leaves, and that neck length increased in a living giraffe as it kept reaching higher and higher for food. This increase of neck length could then be passed on to the next generation. Today, evidence suggests he wasn't completely right about how adaptation occurs, and many texts and articles downplay Lamarck's contributions to science and portray him as the guy who had it wrong before Darwin came and corrected him.

CHARLES DARWIN

Born in 1809, Charles Darwin (Figure 4.6) was raised by an intellectual and progressive family that included doctors and industrialists. By the time he was sixteen, he had gained an appreciation for the natural world,

Figure 4.3: *Brassica oleracea* (A), a wild plant in the mustard family, has been selectively bred to produce many familiar vegetables, including B) Brussels sprouts; C) Cabbage; D) Cauliflower; E) Kale; and F) Broccoli

especially invertebrates, which he thought may be more primitive compared to other creatures. On the advice of his domineering father, he studied medicine at Edinburgh University but dabbled in side projects ranging from geology to taxidermy.

Figure 4.4: Crops such as corn are selectively bred for maximum yields.

Figure 4.5: Jean-Baptiste Pierre Antoine de Monet, Chevalier de Lamarck (1724–1829), a french naturalist, suggested that species acquired traits during their lifetimes that they could pass down to the next generation.

Redirected again by his father, Darwin was sent to study religion at Cambridge as a way to balance out his obsession with nature. Ironically, it was this change of course that led Darwin to later conceive his most famous theories. At Cambridge, in his early twenties, Darwin met Reverend John Henslow, a botanist who was interested in the natural world himself and became a mentor to Darwin. It was Henslow that told Darwin about a ship, the HMS *Beagle*, that would be piloted around South America by an evangelist named Robert Fitzroy whose mission was to map parts of the South American coastline. This gave Darwin the opportunity to observe and collect many exotic species of plants and animals.

The five-year voyage (1831–36) on the *Beagle* was an adventurous one for Darwin. He helped rebels in Uruguay retake a fort and experienced a catastrophic earthquake and tsunami in Valdivia, Chile. He collected fossils such as giant skulls of extinct mammals and contacted tribes who had never been introduced to any inkling of the Western world. His experiences left him with a new sense of purpose, and he spent the next decade formulating his theory of evolution (at the time called descent). Surprisingly, Darwin had initially thought the

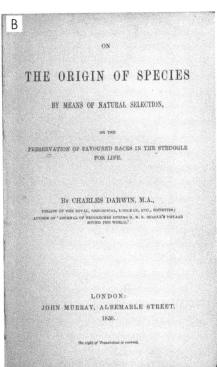

Figure 4.6: A) Charles Darwin (1809–1882); B) The front cover of the 1859 edition of Darwin's *On the Origin of Species*.

variety of Galapagos finches collected were completely different types of birds. It was a colleague of his, John Gould, from the local zoological society that convinced Darwin they were variations of a more similar group of finches.

Meanwhile, he supposedly lost his faith in God after several of his children died early from diseases such as typhoid. He moved to a remote town about sixteen miles from London and became somewhat of a recluse, apprehensive about publishing his work for fear of the controversy it would create. Although he put into writing a formal description of his theory in 1844, he waited until the late 1850s to publish it, as by that time science had become more secular and freer from the threat of religious persecution. His famous work *Origin of Species* was first published in 1859, although the term "evolution" didn't show up until the last edition published in 1872.

ALFRED RUSSEL WALLACE

Alfred Russel Wallace, a contemporary of Darwin, was born in 1823 (Figure 4.7). He had a poor childhood and worked as a railway surveyor in his teenage years. He was a self-taught naturalist, and at the age of nineteen, he happened to befriend a young man named Henry W. Bates, who was a beetle collector and a source of inspiration to Wallace. Wallace and Bates continued to be fascinated by insects and decided in 1848 to travel to Brazil to study exotic species of plants and animals. While in South America, Wallace split away from Bates, who had slightly dissimilar scientific interests.

By 1852, Wallace was heading back to England on a boat filled with years of collected specimens and notes. Unfortunately, while crossing the Atlantic, his boat caught fire and his collection burned and sank to the bottom of the sea. Luckily, he was able to salvage his sketches, and from these, he wrote his first theoretical paper about evolution, which he described as the mechanism of descent with modification. He also suggested that all living things

Figure 4.7: Alfred Russel Wallace at age 24.

are somehow related. Not long after, Wallace published these writings in a paper titled "On the Law Which has Regulated the Introduction of New Species."

In the 1858 annual meeting of the Linnean Society of London, the theory of evolution was introduced as the brain-children of both Wallace and Darwin. It just so happened that in later years, Darwin gained popularity, while Wallace, who at the time of his death was one of the most noteworthy scientists of his day, became overshadowed.

GREGOR MENDEL

Early scientific research of inheritable traits began in the mid-nineteenth century with a set of observations described by an Austrian monk named Gregor Mendel (1822–1884) (Figure 4.8). Mendel's work wasn't appreciated much for its scientific merit until decades after his death, when the modern study of genetics took form. Today, his research stands as a widely used example of the natural laws that govern the passing along of traits to subsequent generations. His interest in plant hybridization gave him a means of demonstrating how traits are inherited, although during his lifetime, his ideas did not become widely accepted as they are today.

Mendel's laws of inheritance differed from the standard view of "blending theory," which stated that specific traits of offspring are the result of the blending of traits acquired from both the parents. Mendel showed that traits don't necessary blend, but rather compete for dominance. This phenomenon is demonstrated well in Mendel's famous study of pea plants (Figure 4.8B).

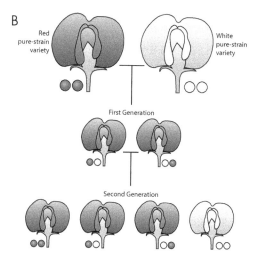

Figure 4.8: A) Gregor Mendel (1822–1884), who helped establish the laws of inheritance; B) Gregor Mendel's experiment with cross-breeding pure-bred strains of pea plants showed that traits such as flower color do not blend in subsequent generations. When red-flowering pea plants and white-flowering pea plants were cross-bred, the first generation only exhibited red flowers. However, when the first generation was inbred, a second generation included some plants with white flowers.

SPECIATION

Speciation is the formation of a new species from an ancestral one. In this book, "ancestral" can also refer to a species coexisting with its newer forms. **Allopatric speciation** occurs when a subset of a species' population becomes isolated by some physical or reproductive barrier and is cut off from the gene flow of the rest the population. Over time, the physical differences that accumulate in this new population make them different enough from the original that they become incapable of interbreeding with them and are at this point a new species. When a species gives rise to multiple new species that vary in their respective adaptations, it is called **adaptive radiation**. This is further explained in the next chapter.

PHYLOGENETIC TREES AND RATES OF SPECIATION

The traditional model of evolution states that organisms adapt to their environments over time and that speciation is a gradual process. This phenomenon can be illustrated with a **phylogenetic tree** (Figure 4.9). which is a type of graph that shows relationships between organisms today and their ancestors.

The theory of evolution as proposed by Darwin predicts that when a new species is formed, it initially appears very similar to the ancestral species it branches from. Divergence of two species has traditionally been thought to occur at a slow, steady rate, so that over time a species and its ancestral stock become progressively more different. Moreover, the standard view of **natural selection** suggests that a species is always changing and adapting during its reign on Earth. This is referred to as **phyletic gradualism**.

Observations of the fossil record suggest a different mechanism of natural selection has *also* been at work than that proposed by Darwin and Wallace over 150 years ago. Evolutionary theory made predictions about the fossil record before much of it had been studied. One of these predictions was that new species diverge gradually from their ancestral stock. Since Darwin's time, scientists have unearthed a plethora of new fossils that have helped us piece together a physical record of evolution. We expected to see species gradually and continuously change throughout their existence, as well as plenty of transitional species marking the growth of new branches on the "tree of life," as phyletic gradualism suggests. Instead, there are three observations of the fossil record that challenge this idea:

1. Many species in the fossil record don't change drastically during their existence. Rather, most of the changes we see occur during rapid speciation.

2. We don't see many transitional organisms marking the development of new species. Species appear rather suddenly in the fossil record, then don't change very drastically after that. Furthermore, new species and their ancestral stock often overlap in time, coexisting for long periods.

3. There are two brief episodes in the history of life that involved profound increases in species diversity and the development of new phyla. The first was the Cambrian Explosion and the second was the invasion of inland areas during the Silurian.

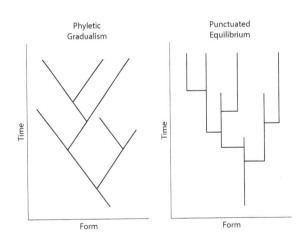

Figure 4.9: Two distinct models of the rate of evolution throughout history.

By the early 1970s, a new theory about natural selection was proposed that better explained these observations stated above. The term **punctuated equilibrium** was coined in 1972 by Niles Eldredge and Stephen Gould of the American Museum of Natural History. They had noticed that many species show **stasis** during their existence; in other words, any changes in a species over time tend to hover about, but don't deviate far from, some ideal design. It's as if there is some kind of resistance toward adapting to a new environment. How, then, does natural selection work in the contexts of phyletic gradualism and punctuated equilibrium?

NATURAL SELECTION

For scientists who believe that phyletic gradualism is the norm, natural selection is the response of organisms to environmental pressures. Variation within a species allows some of its members to be better equipped to overcome these pressures and ultimately pass on these traits to new generations. Evolution, then, is a response of a species to a changing environment through constant adaptation. The general lack of transitional fossils representing slow, progressive speciation events is explained as a product of unconformities and other natural phenomena that erase parts of the fossil record.

Indeed, environments are dynamic and undergo changes through time from the perspective of a stationary location, but that doesn't mean an environment can't shift to a new location without completely going away.

For scientists who believe that punctuated equilibrium better defines the pattern of evolution, natural selection is seen more as a process where organisms determine their environment, not the other way around. Climate change and plate movement cause environments to shift their geographic locations, and it is common to see the inhabitants of these particular niches follow. Many of those who support punctuated equilibrium also tend to believe that organisms seek environments that suit their lifestyles and respond to environmental change by moving to a new spot that most closely matches the environment they are already adapted to. This can allow a species to remain static in an evolutionary sense as long as it can keep up with a spatially shifting ecosystem. For instance, throughout the Pleistocene epoch when the climate repeatedly warmed, belts of Alpine forests including oak, cottonwood, and aspen repeatedly shifted northward into Canada as well as to higher elevations in the Rockies. These trees still exist today despite having to relocate many times.

So how, then, does speciation take place if organisms remain relatively static? The answer is pretty much the same as it is for Darwinists: Even a species that remains static for millions of years can yield a variety of offshoots, especially around the periphery of their population, where isolation most easily occurs. If a species is removed from its environment and *cannot* seek out a replacement with similar conditions, it is forced to adapt or go extinct. Adaptation occurs more quickly in smaller groups with more restricted **gene** pools, resulting in the seemingly rapid-fire nature of speciation shown in places by the fossil record.

Typically, each environmental niche is occupied by one successful species, with the other offshoots lying in wait for the right opportunity to take over. When one species goes extinct, the next closest tends to take its place. This can lead to a macroevolutionary pattern determined by differences *between* successive species and not within the time span of only one. Furthermore, major extinction events are typically followed by periods of intense speciation and diversification. Apparently, the vacancies left behind in ecosystems after extinction events are filled quickly by newer species that had been lying in wait for their ancestral stock to disappear.

EXAMPLES OF NATURAL SELECTION

PEPPERED MOTH

A major driving force of natural selection is the influence of predators. The peppered moth is a perfect example of short-term evolution driven by the selective force of predation. In many forests of Britain and North America, a moth called *Biston betularia* rests on the bark-covered trunks of birch trees. The normal color of birch bark is light gray with dark speckles, partly colored by lichen. The peppered moth got its name because it had adapted to its environment by developing wings that camouflaged against the light-colored, speckled lichen and hid them from predatory birds.

As with any species, natural variation results in small differences within a species' population. A genetic mutation causing the wings of the peppered moth to be dark gray in color had sprung up in the species, but it tended to be weeded out through predation. The dark gray moths didn't camouflage well against the light bark, and they were most easily found by predators. This mutation was not advantageous until conditions surrounding urban centers changed during the Industrial Revolution.

When industrialization took place, many smokestacks sprang up and began to belch dark soot into the air. This soot collected on the bark of birch trees nearby, killing the lichen and making it much darker in color than it had been before. Under these new circumstances, the dark-wing mutation became advantageous, as those moths lucky enough to develop dark gray wings blended in well with the dark tree trunks. Over several decades, the populations of peppered moths became dominantly dark-colored because predatory birds tended to miss them amongst the soot.

As smokestacks got taller and progressively less soot was emitted, the birch trees showed their natural light color again and the melanism that had been so beneficial to the moth during the industrial period became disfavored yet again.

DESERT POCKET MOUSE

In the arid and rocky southwestern United States and northern Mexico, there lives the desert pocket mouse, another example of natural selection at work (Figure 4.10). The desert floor in this region is typically composed of light tan to light gray rocks to which the pocket mouse has adapted camouflage. Therefore, most desert pocket mice have light tan fur that helps them avoid predators such as birds, reptiles, and predatory mammals by blending in.

Dotting this landscape are dark, almost black rocks derived from basalt flows, some of which are as young as a thousand years old. The pocket mice that live within these black rocks have developed black fur. The black fur comes from a mutation in the gene controlling fur color and pattern. Statistically, only a few out of a hundred thousand mice undergo this mutation, and in the light rocks, they are easily picked out by predators.

The light tan pocket mice, however, contrast greatly with the dark basalt, making *them* stand out to predators instead. Today, the pocket mouse populations that live in the dark rocks are predominantly dark-furred.

The mutation that made them stand out so well on the tan rocks makes them perfect for hiding out in the dark rocks, and those genes became favored over time. Consequently, nearly all the mice that live in the dark basaltic rocks have dark fur, and only tens of meters away, in the light-colored rocks, the light

Figure 4.10: The desert pocket mouse (*Chaetodipus penicillatus*).

tan mice prevail. This dichotomy of color exists in many places throughout this region, making it unlikely this was just an evolutionary coincidence in just one group of mice.

EVIDENCE OF EVOLUTION

The evidence of evolution is all around us and even part of us. The predictions made by evolutionary theory have more or less been confirmed by the fossil record.

THE FOSSIL RECORD

The fossil record consists of the cumulative discoveries of all fossilized life discovered thus far. Humans and their ancestors have been discovering fossils for millions of years but didn't know what to make of them. The ancient Egyptians that constructed the pyramids, for example, found fossils of bean-shaped Foraminifera in the limestone blocks used in construction. It is rumored that the slaves who built the pyramids saw these fossils, which are about the size of dimes, and interpreted them to be the petrified remains of lentils eaten by previous generations of builders. Given the abundance of these foraminifera tests in the limestone, these slaves must have thought the earlier workers wasted a lot of beans.

For thousands of years, naturalists and philosophers from around the planet continued to discover fossils and interpret them as the preserved remains of older forms of life (Figure 4.11). Particularly since the Middle Ages, these fossils were thought to have been derived from great biblical floods.

Leonardo da Vinci (1452–1519) (Figure 4.12) did not subscribe to the notions of the day that fossil clams found on the tops of mountains were the result of a worldwide (biblical) flood, a long-standing idea that continues with much of the world's population today. These theories suggested the fossil clams just grew by themselves in the rocks. Da Vinci used his critical thinking skills to address the fallacious nature of these claims. Here are some of the reasons why da Vinci thought "such an opinion cannot exist in a brain of much reason" citing several facts:

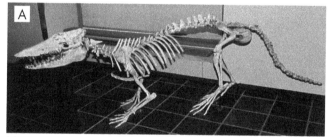

Pakicetus (Early Eocene; 50+ million years ago)

Ambulocetus (49 million years ago)

Kutchicetus (45 million years ago)

Figure 4.11: The evolution of early whales is well-documented in the fossil record.

1. Clams are marine organisms, and it's simply illogical that the mountaintop clams, with growth rings and evidence of articulation, could ever have formed that way in solid rock.

2. If there was a great flood that brought these fossils to the mountaintops, then where did all this water go?

3. Water washes debris downhill, not uphill.

4. A large deluge of water would not leave behind such clean and well-organized sedimentary layers.

Logic is the enemy of dogma, and his theories are still looked upon today by many people as heretical.

Figure 4.12: A sketch called *A Rocky Ravine* created by Leonardo Da Vinci, showing his keen interest in geology.

Naturalists such as Nicolas Steno and John Woodward (1665–1728) recognized that geological strata, which contain these fossils, stack up on top of each other, and thusly, one could assume that deeper layers contain older fossils. What was still missing then was a scientific interpretation of the ages of these layers. Did the fossil organisms live and die in the six thousand years of Earth history claimed by Young Earth theorists? A more detailed examination of the rates of geologic change was needed to provide a more evidence-based idea of the age of the Earth, and therefore the true scope of the fossil record.

James Hutton helped to contradict a young Earth by introducing uniformitarianism, the notion that geological processes such as erosion and deposition are slow, ongoing phenomena, and thusly, a stack of rocks thousands of feet tall incised by canyons hundreds of feet deep would require tremendous amounts of time, at least millions of years, to form.

With the plethora of evidence today suggesting the Earth is billions of years old and the recognition of fossils in rocks from throughout much of this period, the fossil record is now viewed as not only a testament to faunal succession, but also of an old Earth with a rich and long-lasting history of life and evolution.

One must recognize, however, that the fossil record consists of what humans have discovered while simply scraping the surface of our planet. Missing links aren't proof that evolutionary theory has flaws; they are only placeholders for future discoveries. As new fossils are constantly being discovered, the fossil record becomes more and more of a continuum.

HOMOLOGOUS STRUCTURES

Homologous structures are physical traits that appear similar in organisms that are presumed to have a common ancestor (Figure 4.13). Vertebrates, for instance, likely share a common fish ancestor (see Chapter 8).

Superficially, the forearm of a human, forelimb of a horse, and wing of a bird appear to be vastly different types of appendages. The bone, muscle, circulatory, and nerve structures, however, are very similar. In fact, all vertebrates share essentially the same general anatomy. The differences lie in the relative lengths and shapes of bones. Our human forearms are constructed of a humerus (attached to the shoulder), a pair of bones (radius and ulna) in the mid-arm, and a set of wrist bones ending with finger bones. The exact same skeletal architecture exists in the forearm of a lizard, a salamander, and a dog, for example, with only minor differences in bone size and length

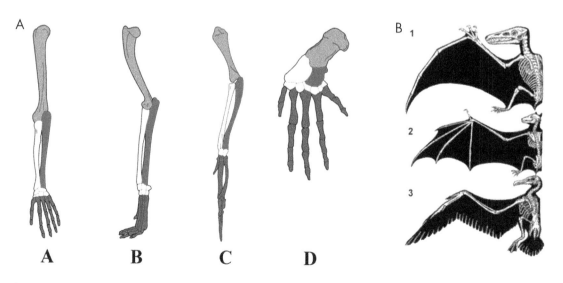

Figure 4.13: A) Homologous forelimbs in a variety of vertebrates; B) Homologous wing structures.

ratios. The same bones are also found in the wings of a bat, whose finger bones have become more elongated to provide a rigid but foldable framework for the wing. Birds, too, have these same bones, except the finger bones have fused together. The leg of a horse is constructed the same way, except only the middle finger remains as a hoof, and the others have become reduced and line the ankle bone.

Another example of skeletal homology is that almost every vertebrate has seven vertebral bones in their necks. We have the same number of neck bones as a giraffe, whose vertebrae have become elongated, and also the whale, whose vertebrae have been reduced to tiny bones that provide little, if any, flexibility. Homology is not limited to vertebrates, however. Many organisms with common ancestors share homologous structures such as the lophophore found in bryozoans and brachiopods (see Chapter 8).

Land plants have a common, moss-like ancestor that had tiny leaves. Over the past nearly half a billion years, plants have evolved with leaves that serve many different functions besides collecting light for photosynthesis. Some plants are carnivorous and use their leaves for feeding. The pitcher plant, for instance, has a cup-shaped leaf that traps insects that fall into it and can't climb out. The Venus flytrap has developed touch-sensitive leaves that close onto an insect like a bear trap. The cactus has developed leaves that act as protective spikes, making it harder for other organisms to tap into the stored water supply in their stems. Some tropical leaves can be over five feet across, while pine needles tend to be skinny and somewhat inflexible. Leaves of the oleander are extremely poisonous, so don't eat them.

VESTIGIAL STRUCTURES

As organisms evolve, some traits may not be needed anymore and begin to reduce in size and function. These are called **vestigial structures** (Figure 4.14). If evolution is descent with modification, then the decrease in the usefulness of particular derived traits indicates that species change over time, and thus evolve.

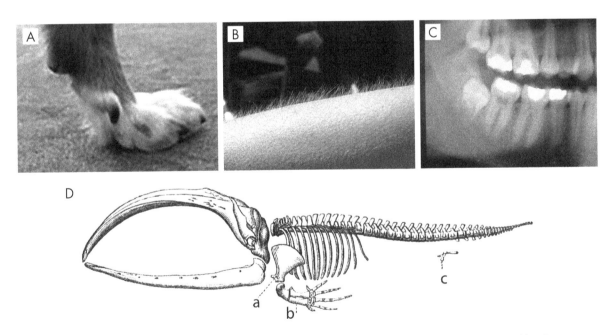

Figure 4.14: Examples of vestigial structures; A) The dewclaw of a dog; B) Goose bumps, caused by the arrector pili muscles; C) X-Ray of wisdom teeth that don't have enough room to grow; D) A whale skeleton including the "free-floating" femur and pelvis left behind by ancestral back legs

For instance, the direct ancestors to wolves (and dogs) had five toes, all of which contacted the ground while walking. Today, dogs still have five toes, but only four contact the ground. The fifth digit (equivalent to our thumb and/or big toe) is now present as the dog's dewclaw, that little stubby toe-like thing partway up their leg that seems to have no purpose. In fact, many vertebrates have vestigial toe bones, even the horse, whose other four digits have shortened and become part of the ankle area.

Whales have no back legs or hind flippers, yet they still have a pelvis and femur, which in modern whales are embedded in their flesh and not attached anymore to the rest of the skeleton. The presence of a vestigial pelvis and femur in the whale implies it had a land-based ancestor; otherwise, these bones have no context.

We humans have several body parts that can be considered vestigial. For example, our jaws have become shorter as we evolved, and consequently, our *wisdom teeth* often cannot fit and must be removed surgically. Thirty percent of people don't even get their wisdom teeth anymore. Our *appendix* is an organ that we consider to have very little use; it may be the remnant of a more complex part of our ancestors' digestive system used for processing hard-to-digest plant matter.

The mammal-like reptiles we believe we are descended from had several extra bones in their jaw. As mammals continued to evolve, these bones shrank and became the *incus* and *malleus* of our inner ear.

Our tree-dwelling ancestors had tails for grasping and keeping balance. We still have tail bones, but they have reduced and fused together into our *coccyx*, or tailbone. When organisms get cold, they puff up their fur or feathers to trap more air to keep them warm. When we get cold, tiny muscle fibers in our skin causes our hairs to rise (goose bumps), but since we have such little hair, it doesn't do us much good. Even the corners of our eyes retain vestigial third eyelids, which are still used in many other organisms as extra eye protection.

EMBRYONIC DEVELOPMENT

Amphibians, reptiles, birds, and mammals display striking similarities of internal anatomy, as discussed earlier in the context of homologous structures. These major groups of animals, as well as fish, their common ancestor, are even more identical while in their embryonic stages (Figure 4.15). During early stages of growth, virtually all vertebrates appear nearly indistinguishable from each other.

For each different organism, genes are predestined to switch on and off during the embryonic stage in a specific pattern that results in the early stages of the adult form. This period of genetic coding, for example, allows for particular body parts, such as wings, fins, flippers, or arms, to develop from the same initial limb bud on all early vertebrate embryos.

Amongst the samples Darwin brought back from the Galapagos were finch eggs containing embryos in particular stages of development. It was determined that the beak shape of the adult finch is influenced by *exactly when* the embryonic genes kicked in to begin growing the beak. Between some finches, the difference was only a few hours and resulted in very different beak shapes.

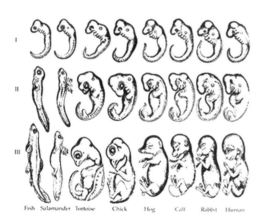

Fish Salamander Tortoise Chick Hog Calf Rabbit Human

Figure 4.15: Sketch by G.J. Romanes after an original diagram by Ernst Haeckel of the similar embryonic development of several vertebrates. Some have criticized these sketches as being exaggerated.

MICROEVOLUTION

Organisms with short generational time spans have the tendency to evolve more rapidly than longer-living organisms.

Bacteria, for example, multiply at extremely fast rates; some species of hydrothermal vents, for example, can double their population in ten minutes. Bacteria and viruses that make us sick evolve so rapidly that as soon as we come up with medicine to fight them, they begin to gain resistance almost immediately.

FAUNAL SUCCESSION AND CORRELATION OF STRATA

The fossil record of evolution is revealed through **faunal succession**, the principle that assemblages of fossils succeed one another through time in an order that can be measured.

There is, however, no spot on our planet that contains rocks and fossils from all the time periods. Instead, isolated fragments of the fossil record are scattered across the planet as finite stacks of sedimentary layers. To understand the complete history of faunal succession, we must stitch together these different stacks of rocks so that the pattern reveals itself. This is done through **correlation**, the practice of identifying rocks from the same time period through the matching up of lithology and/or fossils (Figure 4.16).

As discussed in Chapter 2, overlaps of sedimentary sequences across the globe are typical. When the upper part of Sedgwick's Cambrian series of rocks correlated with the lower part of Murchison's Silurian series, the overlap was given the name Ordovician and it became its own series. The correlation was based on the similarity of fossil assemblages in the overlapping stacks. There are two main ways in which we can correlate rocks.

Lithostratigraphic correlation is the practice of identifying rocks of the same general lithology and, presumably, the same age. In other words, we can correlate, in a lithostratigraphic sense, two different bodies of rock if they have similar compositions. Without fossils, it is difficult to establish time equivalence with just the rock type alone. It helps if the rock is a distinctive layer or set of layers that are laterally extensive and contain unique markers that can be matched up across large distances. This type of correlation can be difficult, as even a single layer of rock from a particular age may look very different depending on the location where it is described.

It is most helpful to correlate layers of volcanic ash in stacks of rock if there are no fossils to help establish time equivalence. Volcanic ash and lava flows are technically igneous, but they stack up with other sedimentary layers according to the laws of superposition and can be directly dated using absolute dating methods. If a layer of ash or lava flow could be correlated, then so can the sedimentary rocks above and below them, assuming there are no unconformities.

A useful method that employs both relative *and* absolute dating is called **bracketing** (Figure 4.17). This can be done when two or more igneous layers can be dated radiometrically, and the sedimentary layers between them can be dated relative to the established ages.

Geological systems are distinct sets of rocks that can be subdivided into smaller sets called *formations*, which in turn can be split into distinct *members* and even *beds*. Systems can also be combined to form broader *groups* and *supergroups*.

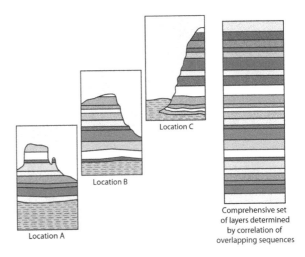

Location C

Location B

Location A

Comprehensive set of layers determined by correlation of overlapping sequences

Figure 4.16: Correlation of sedimentary layers. Three distinct locations may contain sedimentary sequences that, through correlation, are inferred to overlap. The individual sets of layers can then be stitched together to make a more comprehensive geological record.

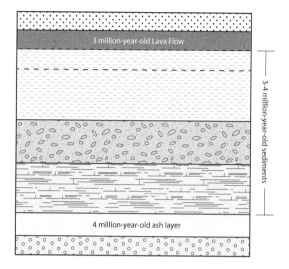

Figure 4:17: Example of "bracketing," wherein the absolute ages of the ash layer and lava flow allow the ages of the intervening beds to be approximated.

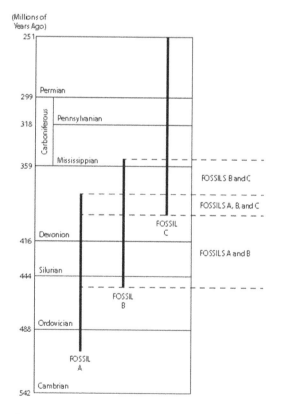

Figure 4.18: Example showing range zones for three different fossils. Although each of the fossils have relatively long range zones, narrower ranges of time mark the periods during which they coexisted. Thus, by looking at fossil assemblages, we can determine the ages of rocks more precisely.

Biostratigraphic correlation is a better way to establish time equivalence because it focuses on fossil content and not lithology. For instance, two different outcrops of sandstone cannot easily be correlated with time equivalence based on lithology alone; sandstone has been forming for billions of years. On the other hand, if the two sandstones have similar fossil assemblages, time equivalence can be established much more easily.

The basic units of biostratigraphy are **biozones**, which are a range of sedimentary strata defined by the unique sets of fossils they contain. Biozones are not characterized by their lithology (like formations), and they may or may not line up with lithostratigraphic units.

Of course, fossil assemblages rarely appear and disappear all at once. Ecosystems change over time, as do their inhabitants. Fossil assemblages are mere snapshots of earlier times, and the way they are differentiated from each other depends on the fossils being considered in each biozone.

A common type of biozone is a **range zone**, which is a span of time in which a particular fossil or assemblage of fossils lived (Figure 4.18). Every species begins at speciation and eventually ends with extinction. Many of the fossils we observe are from organisms that went extinct many eons ago, and thus we can use them as **guide fossils** for the period of time during which they existed.

Some organisms make better guide fossils than others. The most useful guide fossils were around for short periods of time but were numerous and widespread during their existence. Stromatolites, for example, can be found fossilized in rocks throughout the world, but they do not make good guide fossils because they have been around for 3.5 billion years (Figure 4.19). Archaeocyathids, however, were sponges that existed only in the Lower Cambrian and were around for only twenty million years, making them excellent guide fossils.

Even two guide fossils that were both around for long periods of time can provide good dating accuracy if they only overlapped for a short period. Consider that cars have been commonly used since the early 1920s, and before that, horse-drawn buggies were the normal mode of transportation in the streets. The period between 1900 and 1920 saw a mixture of horse-drawn buggies and early automobiles. Although either mode of transportation has existed for a long range of time, both were common on the streets for a rather short period.

Defining the overlap of range zones between two or more fossils helps to establish **concurrent range zones**, periods of time when fossils of slightly different range zones coexisted. Establishing concurrent range zones is one of the best methods of determining time equivalence between rocks separated by large distances.

Stromatolites
(3.5 billion years ago - Present)

Rudist Bivalves
(Late Jurassic - Late Cretaceous)

Figure 4.21) The stromatolite on the left has a much longer range zone than the rudist bivalve on the right.

Figure 4.19: A) The stromatolite on the left has a much longer range zone than the rudist bivalve on the right. B) Rudist bivalves from the Cretaceous, Omani Mountains, United Arab Emirates.

5

Figure 5.1: A "liger," the offspring of a lion and a tiger, which has inherited traits of both parents.

CLASSIFICATION

DNA[1]

POTASSIUM, ARGON, DNA, AND WALKING UPRIGHT

From a strictly biological perspective, humans are not that different from chimpanzees or gorillas. We all have countless anatomical features in common—including opposable thumbs and fingernails—and our genetic blueprints differ only by a few percent. However, the small number of biological traits that do distinguish us from other apes correspond to enormous differences in behavior. A chimpanzee, after all, is unlikely to write a book like this one, and even if one did, it would have great difficulty finding a publisher. Therefore, if we can determine the origin of those characteristics unique to humankind, we can better understand and appreciate what it is that makes us special.

Two of the most obvious physical traits that distinguish humans from other great apes are our greatly enlarged brains and our style of walking on two legs with the torso held vertically. Our increased brain size clearly has a direct relationship to our uniquely complex behavior and culture. However, our posture also appears to have played a pivotal role in our evolution. Our ancestors walked on two legs long before they began to have bigger brains, and among all the ancestors of humans and great apes, only the bipedal creatures demonstrate dramatic increases in brain size. It is therefore possible that this peculiar mode of locomotion somehow facilitated later changes in brain structure, although the details of this relationship are far from clear.

The origin of our bipedalism is also a hot topic for anthropologists because it is a longstanding puzzle that recent discoveries may finally help solve. Changes in walking style, like changes in brain size, involve alterations of skeletal features, so in principle the fossil record should provide important information about when, where, and how both these crucial adaptations occurred. However, while ancient skulls from Africa document changes in brain size

---

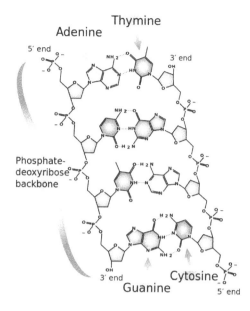

Figure 5.2: Structure of the DNA molecule.

among our ancestors over the past five million years, no one has yet found old bones that clearly indicate when or where our ancestors began to walk upright. This lack of detailed information about the circumstances surrounding the origin of bipedalism has made it very difficult to determine why walking on two legs became advantageous for our ancestors. But in the last decade, teams of researchers working in Ethiopia, Kenya, and Chad have uncovered some very interesting fossils. At present, the recovered material is still rather fragmentary, but the dates associated with these finds are enough to make them extremely exciting to anthropologists. The ages of these newly discovered bones indicate these early bipeds lived at a time that—according to DNA evidence—may have been a crucial turning point in the history of our lineage, so these creatures may document the earliest stages of the unique traits like bipedalism that made us what we are today (Figure 5.2).

THE HOMINIDS

The newly discovered fossils—like most fossils that anthropologists use to study the origins of our uniquely human traits—belong to a set of animals that all have some of the traits that make today's humans unique, such as enlarged brain size and bipedalism, as well as a host of subtler features including reduced canine teeth and barrel-shaped rib cages. These creatures used to be referred to as hominids, but thanks to recent refinements in the classification system they are now often called hominins instead. I will use the older, more familiar term here, but regardless of what

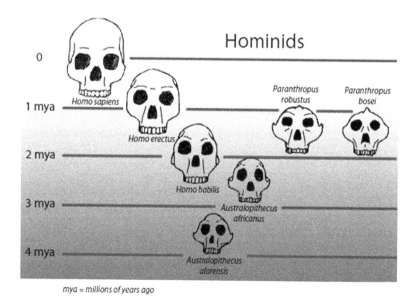

mya = millions of years ago

Figure 5.3: Drawings of various hominid skulls, all to scale. The position of the skulls along the vertical axis indicates the age of the fossil, which the horizontal axis is arbitrary. Increases in brain size can be observed as the dome of the skull rises over the top of the brow-line.

you name them, the combinations of characteristics they share with us are unlikely to appear multiple times in different animals, so the hominids and modern humans almost certainly inherited these features from a common ancestor. The distribution of traits among hominid fossils can therefore reveal how and when our ancestors acquired these characteristics.

For example, consider brain size. Figure 5.3 shows the skulls of several different kinds of hominids from different times. The amount of skull rising over the eyebrow ridge provides a rough indication of brain size, so we can clearly see that the earliest hominids like *Australopithecus afarensis* have relatively small brains—comparable to those of modern chimpanzees. We can also observe a definite trend: brain size increases from *Australopithecus afarensis* to *Homo habilis* to *Homo erectus*, and finally to modern humans (also known as *Homo sapiens*). However, we can also see some hominids did not follow this same trend. For example, the brain size of the *Paranthropus robustus* is considerably less than that of the contemporary species *Homo erectus*. By comparing the habitats and diets of these different hominids, paleoanthropologists are able to gain insights into the processes that fostered the evolution of large brains. The data also indicate that hominids with enlarged brains began to appear about two million years ago, so the relevance of any climatic trends or other environmental phenomena at this time can be explored.

By contrast, all well-preserved, nearly complete hominid fossil specimens—even those from *Australopithecus afarensis*—have features that indicate these creatures walked on two legs: their lower spine was curved backwards in order to support a vertical trunk and their hip, knee, and ankle joints allowed their legs to swing forward and backward under the pelvis. This means that the ability to walk on two legs is not only older than enlarged brains, it is also older than *Australopithecus afarensis* or any of the other hominids shown in Figure 5.3. However, until someone discovers fossil hominids that could not walk efficiently on two legs, it will be very difficult to ascertain what caused our ancestors to adopt this mode of locomotion. Anthropologists seeking the origins of bipedalism have therefore been searching for fossil hominids predating *Australopithecus afarensis*, and these efforts have recently started to be rewarded.

In 2001, teams of paleoanthropologists working in Ethiopia and Kenya announced that they had found fragmentary remains of hominids. While only a few bones were found at each location, the characteristics of the teeth were sufficient to distinguish them from *Australopithecus* and other hominids, so they were given the names *Ardipithecus ramidus* and *Orrorin tugenensis*, respectively. More recently, a team digging in Chad found a well-preserved skull of yet another hominid, which they called *Sahelanthropus tchadensis*. These bones all appear to be older than any previously known hominid remains, and therefore document a previously unexplored period of hominid history. As with the early American sites described in the last chapter, the evidence for the antiquity of these new finds comes from a radiometric dating method based on an unstable isotope. However, in this case the isotope is not a form of carbon created high in the sky, but a form of potassium released from deep underground.

POTASSIUM-ARGON DATING AND THE AGE OF HOMINID FOSSILS

Potassium atoms all have nineteen protons, and depending on the isotope, they can have varying numbers of neutrons. Most potassium on Earth is in the form of potassium-39, which has twenty neutrons and is completely stable. However, about 0.01% of potassium atoms are in the form of potassium-40, which has twenty-one neutrons and is unstable (see Figure 5.4). Like carbon-14, this isotope of potassium can undergo beta decay by having a neutron spontaneously convert into a proton. In this case, the process leaves behind a calcium-40 nucleus. However, 10% of the time potassium-40 decays in a somewhat different way; the nucleus captures an electron, and one of the protons converts into a neutron, producing an atom of argon-40.

Like carbon-14, potassium-40 atoms can be used as timekeepers because they decay with a well-defined half-life determined solely by the number of protons and neutrons they contain. However, while carbon-14 has a half-life of only a few thousand years, potassium-40 has a half-life of 1.28 billion years, so these two isotopes probe very different timescales. If we have material that is some thousands of years old, enough time has gone by for a significant amount of the carbon-14 to have decayed, but very only a tiny fraction of potassium-40 has transformed into calcium or argon. For such comparatively recent material, carbon-14 provides a much more sensitive indicator of age than potassium-40. For objects millions or billions of years old, however, almost all of the

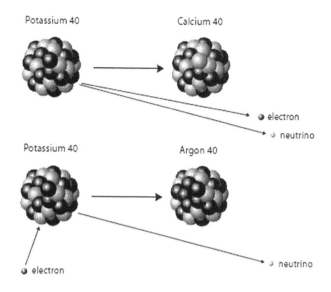

Figure 5.4: Postassium-40 decay. About 90 percent of the time it undergoes beta decay like carbon-14, and a neutron (black circle) converts into a proton (gray circle) to form calcium-40. The remaining 10 percent of the time the nucleus captures an electron, a protron converts into a neutron, and the nucleus converts into argon-40.

original carbon-14 has decayed away while a significant amount of potassium-40 remains. Potassium-40 can therefore be used to measure the age of much older objects.

The types of materials that can be dated with these two methods are also very different because the relevant atoms have distinct chemical properties. As we have already seen, all organisms living at the same time receive comparable amounts of carbon-14 from the carbon dioxide in the atmosphere, so this isotope is often useful for dating material derived from living creatures. By contrast, potassium-40 is best used to date volcanic rocks, not because these all have similar potassium-40 contents, but because all molten rocks ideally contain no argon-40.

Argon belongs to a class of elements called noble gases, which includes both helium and neon. Noble gases are unique in that—except under very extreme circumstances—they do not form chemical bonds with other elements, so the only way they interact with other atoms is by bouncing off of them. Noble gases can easily escape from molten lava because these superheated rocks are in a quasi-liquid state, and their molecules are all moving and jostling past each other. In this environment, the argon atoms can just bounce around until they find the surface and hop off into the atmosphere. By contrast, argon can be trapped in a solid rock because here the atoms are arranged in rigid lattices, which form tiny cages that argon atoms cannot escape from.

Newly formed volcanic rocks ideally should not contain any argon since this gas escaped before the liquid lava cooled into a solid. However, they do typically contain at least some potassium-40. As time goes on, this potassium decays into argon-40, which remains trapped in the solid rock and allows us to estimate how much potassium-40 has decayed since the rock was formed. This data, combined with the current potassium-40 content of the rock, gives us all the information we need to compute the age of the rock.

For example, suppose we find a volcanic rock that currently contains 10 micrograms of potassium-40 and 1 microgram of argon-40. This means that 1 microgram of potassium-40 has transformed into argon-40 during the time since the rock first formed. Since only about 10% of potassium-40 atoms transform into argon-40, we can conclude that a total of 10 micrograms of potassium-40 has decayed over the course of the rock's existence. The rock therefore originally contained 20 micrograms of potassium-40 when it first cooled out of the lava flow, which means that one-half of the original potassium-40 atoms have decayed by the present day. This tells us that the rock must have solidified one potassium-40 half-life, or about 1.28 billion years, ago.

This method of measuring the age of volcanic rocks—called potassium-argon dating—is a simple and elegant way of measuring age because we can deduce the original potassium-40 content of the rock directly from the materials in the rock, and we do not need to estimate it through additional calibration data. In other words, the age estimate relies only on data provided by the rock itself and the only assumption made is that the rock initially contained no argon-40 at all. Still, this technique is certainly not foolproof, as there are a variety of processes that can contaminate the argon-40 content of a rock and corrupt the age estimate. The original lava flow may have contained some unmelted rocks, raising the argon-40 content of the lava above zero, or the rock may have been heated sometime after it formed, allowing some of the argon-40 to escape.

Just as with carbon-14 dating, there is a mini-industry dedicated to refining this technique and to developing procedures that provide accurate and reliable ages. For example, the reliability of potassium-argon dates can be evaluated using a clever bit of alchemy. Prior to extracting the argon, scientists can bombard the rock with neutrons from a nuclear reactor. Just as neutrons from cosmic rays convert nitrogen-14 into carbon-14 in the upper atmosphere, these neutrons transform some of the potassium-39 in the rock into argon-39. Since most of the potassium in any rock is in the form of potassium-39, this process generates a form of argon that "traces" the potassium content of the rock. After this treatment, the rock is gradually heated in order to extract the argon, and the argon-39 and argon-40 are separated and measured using mass spectrometry. Since both argon-39 and argon-40 were produced from the same element, the ratio of argon-39 to argon-40 should be the same throughout the rock. However, if the rock had been heated sometime earlier in its life and some of the argon-40 escaped, different regions or minerals in the rock will have discordant mixes of argon-39 and argon-40. Comparing the ratios of argon isotopes in the gas extracted from the rock at various temperatures therefore provides a way to determine whether the potassium-argon age has been corrupted.

VOLCANOES AND FOSSILS IN EAST AFRICA

Potassium-argon dating allows us to determine when volcanic rocks first solidified. Since bones do not last long in the heat of a lava flow, we are not likely to find contemporary fossils embedded in these sorts of rocks. Most fossils are instead found in sedimentary deposits, where layer upon layer of mud, sand, or other material was piled up by water or wind. Normally, the potassium-argon method cannot help date sedimentary material directly. However,

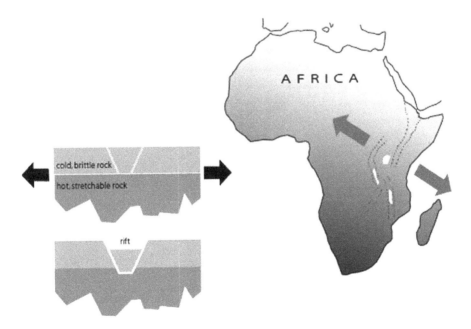

Figure 5.5: The East African Rift System. Left: a cross-section of a rift system. Geological forces pull the crust apart in these locations. The hot lower layer stretches, but the cooler upper layer breaks and a wedge slips down to form a depression. Right: the East African Rift System itself, with the depression indicated by dotted lines. The arrows indicate the forces that may have helped form these features.

thanks to a geological phenomenon called the East African Rift System, the ages of volcanic rocks can help date sedimentary deposits containing early fossil hominids.

The East African Rift System is a place where the earth's crust has been torn apart by processes connected with those forces that caused the various continents to drift across the surface of the globe. These forces are driven by the heat contained deep within the earth, which is so intense that it makes rock pliable, allowing it to stretch and flow. Closer to the surface, rocks are cooler and more brittle, so instead of stretching, the uppermost crust breaks into pieces that slip past each other to form a series of valleys and depressions (see Figure 5.5). This particular rift system extends all the way from Eritrea to Mozambique.

The East African Rift System had three important effects on the local environment. First, water collected at the bottom of the depressions, forming a series of lakes and creating habitats attractive to a variety of wildlife, including some hominids. Second, water and wind carried sediments down from the surrounding highlands into the depressions, which buried and preserved some of the animals as fossils. Third, the stress on and movement of the earth's crust allowed magma to reach the surface, causing widespread volcanic activity that at various times covered parts of this area with ash falls and lava. East Africa therefore contains layers of volcanic deposits interleaved with fossil-bearing sediments.

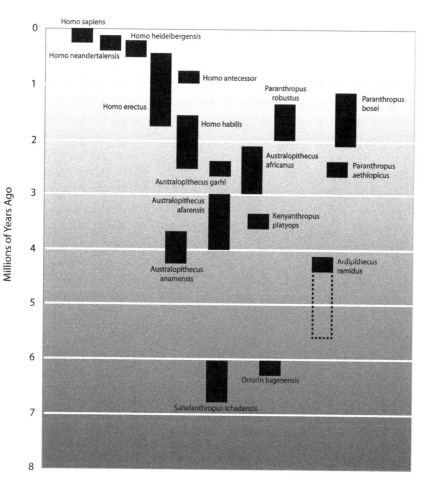

Suppose a fossil-bearing layer of sedimentary rock is sandwiched between two layers of volcanic rocks. The fossil-bearing layer must be younger than the volcanic deposit it sits on and older than the volcanic rocks on top of it, so dating the volcanic layers with the potassium-argon system will provide tight constraints on the age of the fossil-bearing layer. This method has yielded very reliable age estimates for many of the hominid remains from the East African Rift System, including *Ardipithecus* and *Orrorin*. Even hominids found outside East Africa benefit from the rift system age measurements. For example, fossils belonging to *Sahelanthropus tchadensis* come from Chad, hundreds of miles west of the rift system and also far from any volcanic deposits that would facilitate dating. However, these fossils were found associated with the remains of other animals like the wild pig *Nyanzachoerus syrticus*, which are also found in the rift system. Using the method outlined above, paleontologists have deduced that these prehistoric beasts lived

Figure 5.6: Dates of various different types of hominids based on a figure in Bernard Wood's "Hominid Reservations in Chad," *Nature* 418 (2002): 134–136. Bars indicate the range of time the various types of hominids probably lived. The recently discovered hominids *Ardipithecus*, *Orrorin*, and *Sahelanthropus*.

roughly six to seven million years ago. This suggests that the fossils, including *Sahelanthropus*, are from this same time period.

Figure 5.6 shows the ages of the hominids discovered as of 2006. Until about a decade ago, all known hominid remains were from deposits less than four million years old. The newly discovered fossils of *Ardipithecus*, *Orrorin*, and *Sahelanthropus*, however, date back to over six million years ago. These recently uncovered hominids are therefore much older than previous finds, but are they old enough to document the origins of bipedalism? The fragmentary remains include bones from the legs and feet of these creatures, and anthropologists are currently debating what the preserved material tells us about the posture of these early hominids. In spite of this, many anthropologists have high expectations for the material from this time period because studies of the DNA from humans and other primates indicate that these remains derive from a pivotal period in human evolution.

MEASURING RELATIONSHIPS WITH DNA

Like the fossils found within the earth, the DNA within every living thing contains useful information about the history of life. As the methods of molecular biology continue to advance and improve, analyses of these molecules have come to play an increasingly important role in studies of the past. While there is still much work to be done before molecular data can even have a chance of providing age measurements that are as reliable as other methods, recent developments are very promising.

The double helix of deoxyribonucleic acid, or DNA, is a familiar icon in biology. These molecules exist in nearly every cell of our bodies and are composed of two intertwined spiral strands connected by a sequence of base pairs, each one made up of a pair of nucleotides. There are four different nucleotides in DNA: adenine, thymine, cytosine, and guanine, which are usually represented by the letters A, T, C, and G. The sequence of base pairs encodes information a cell needs to function and interact with other cells in a living organism. For example, certain parts of the sequence provide instructions for making different proteins, while other parts determine when these proteins should be made.

The nucleotides have specific chemical properties such that adenine and thymine always pair with one another across strands, as do cytosine and guanine. This means that if one strand has the series ACTTGCT, the other strand must have the sequence TGAACGA. Each of the two strands of the DNA molecule therefore contains essentially the same information. Normally this information is encased inside the coils of the double helix, but the machinery inside our cells can pull these two strands apart as needed so that the information inside can be read. Also, all of the data in a DNA molecule can be replicated by separating the two strands and using each one as a template for the construction of an identical copy of the original molecule. This process occurs naturally every time a cell divides, so that each of the cells maintains a complete set of the instructions required for it to function. By the same token, the information encoded in the DNA of every organism is derived from the DNA of its parents and is inherited by the DNA of its offspring.

Over time, as DNA is passed from generation to generation, the sequence of base pairs changes. These changes are called mutations, and they can occur due to errors in the replication process or because the DNA molecule itself gets damaged in some way. By changing the information encoded in the DNA, these mutations can change how certain cells function and ultimately alter the physical characteristics of the organism. Assuming that the change does not kill the cell or organism, the mutated DNA can be inherited by future generations of cells and organisms. Eventually, this mutated DNA will mutate again, and again that change can be passed on to new cells. As mutations in the DNA accumulate over many generations, creatures descended from a single organism can acquire very different characteristics. Indeed, it is likely that all living things on earth are descended from a common ancestor, and the diversity of life we see now is the result of huge numbers of mutations over billions of years.

Not only is the accumulation of mutations responsible for producing the great diversity of life on earth, it also enables us to uncover relationships between different organisms. Mutations are relatively rare, quasi-random events, and it is improbable—though by no means impossible—that the same mutation will occur twice in different organisms. Furthermore, after a mutation has occurred it is unlikely that an organism with that mutation will revert back to the exact same state as its premutation ancestors. This means that as mutations accumulate, the number of differences between DNA sequences tends to increase with time, and so two organisms with a recent common ancestor will have more similarities than two organisms with a more ancient common ancestor. We can therefore gain insight into the family history of organisms by studying their similarities and differences.

For many decades now, biologists have used the distribution of various physical characteristics to infer relationships between organisms and to investigate the patterns and processes behind the evolution of these traits. Nowadays, thanks to technology that can efficiently read the sequence of nucleotides on DNA molecules, biologists can compare long sequences of nucleotides from different creatures. These new data provide fascinating new insights into the relationships between organisms. In fact, these DNA sequences may provide a new, independent method of estimating how long ago related creatures—such as chimpanzees and humans—began to diverge from one another.

Variations in DNA sequences allow biologists to measure the differences between organisms in a more quantitative way than was previously possible. It is almost impossible to determine whether oak and elm trees are "more different" from each other than dogs and cats based solely on their appearances. Does the fact that cats always land on their feet while dogs don't make them less closely related than elm and oak trees, or do the different shapes of elm leaves and oak leaves mean that there are *more* generations separating them from each other than separate cats and dogs? The difficulties with this sort of inquiry are obvious. Within the DNA molecule, however, all of these differences boil down to the addition, removal, movement, and replacement of a discrete number of nucleotides. Therefore, it is possible to count the number of differences between species on the strands of their DNA. If you wanted, you could see how many times the DNA sequence of a cat had an A where a dog had a T, and then find the number of places an oak has an A where an elm has a T. By comparing these or any of a host of other possible parameters, we could actually make some quantitative statement about the differences between oaks and elms and cats and dogs. This quantitative data is essential for any attempt to estimate ages with biological data from modern animals.

The number of differences between two DNA sequences is a measure of how many mutations have occurred in the two sequences since they diverged from a common ancestor, so we expect that this number will get progressively larger as time goes on. Furthermore, if we assume that mutations accumulate at a steady rate, then this number also is proportional to the time that has elapsed since the two organisms last shared a common ancestor. For example, one stretch of DNA in polar bears differs by 1% from that found in grizzly bears, while the same sequence differs by 3% in wolves and coyotes. If the above assumption holds, we can deduce that the ancestors of the wolves and coyotes have had three times as long to accumulate mutations, so if polar bears and grizzly bears last shared a common ancestor about half a million years ago, we can estimate that the last common ancestor of wolves and coyotes lived sometime between one and two million years ago.

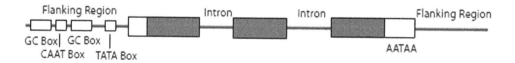

Figure 5.7: The general structure of a gene (based on a figure in Li's *Molecular Genetics*). Only the gray shaded regions contain the information for making the protein. Between these regions are the introns that are ignored when the protein is made.

ATTTCGCTAGCTAGTCGACGACTTCGATCAGCTAGCAGGCATCTGACGAGCT
and
ATATCGCTAGCTAGTCGACGACTTGGAGCAGCTAGCAGGAATCTGATGAGCT

Figure 5.8: Two sequences with a 10% difference.

Living things are much more complicated than nuclear isotopes, so it is reasonable to question whether their mutations could ever accumulate at a fixed or even calculable rate. While it is true that there is still a lot of work that needs to be done before the reliability of this method can be assured, some of the available evidence is encouraging. Many animals use the same basic cellular machinery to read, repair, and copy their DNA molecules, so all of these creatures should be equally prone to mutations. Also, while exposure to certain toxic chemicals or large doses of radiation can greatly accelerate the mutation rate in organisms, such extreme conditions seldom occurred in the wilds of the distant past. We might therefore reasonably expect that mutations would occur at roughly the same rate in all organisms.

Mutations cannot accumulate unless they are passed on to another generation, but often the likelihood of this occurrence involves complex interactions between the organisms and their environment. For example, imagine a mutation that causes a rabbit to have a white coat instead of a dark coat. If the rabbit lived in a forest, it would stand out in the underbrush and promptly get eaten. Few or none of this hapless rabbit's offspring would survive to pass on this mutation. However, if it lived in the arctic, it would be well camouflaged and its offspring would likely thrive. These sorts of mutations will therefore be passed on through the generations at different rates depending on the particular situation. While such variations are of great interest to biologists, these mutations are clearly not going to be the most ideal timekeepers.

Fortunately, there are also mutations that are "silent," which means that they have no discernible impact on the creature's physical appearance or ability to survive. Silent mutations are possible because organisms do not typically use all of the information encoded in their DNA sequences. In fact, while we still do not know exactly what all the information encoded in any creature's DNA means, some mutations can be clearly identified as silent because much of the useful information is contained in segments of DNA with certain recognizable characteristics.

For example, a significant portion of our DNA contains instructions for making various proteins, large molecules made up of long chains of chemicals called amino acids. Proteins are versatile molecules, and different proteins can have very different chemical properties depending on the sequence of amino acids they contain, so proteins are responsible for most of the complex chemical processes that allow a cell or an organism to function. The data required for making proteins are packaged in segments of DNA known as genes. Each gene contains a sequence of nucleotides that encode the sequence of amino acids required to make a particular protein. In order for the cell to be able to translate this information into a functional protein, it needs the flanking regions located on either side of the protein-coding sequence to tell it where the relevant information is. These regions contain characteristic nucleotide sequences that tell the relevant molecular machines where to start and stop reading the DNA.

Similarly, biologists use these flanking regions to identify all of the genes in a given stretch of DNA. It turns out that in humans and other mammals, only a small fraction of the DNA—roughly 5%—is in the form of functional genes. A portion of the remaining DNA still has some utility. For example, there are DNA sequences that are thought to regulate when the various genes are read. However, much of this material can be altered without having any noticeable affect on the cell or the organism. Some of this DNA has even been identified as "broken genes," former genes with mutations in the coding and flanking regions that render these DNA sequences impossible to read. Mutations in these regions should have no appreciable impact on the organism's health or appearance.

Even within the genes themselves, there are regions where changes in the DNA do not affect the structure of the protein. There are stretches of DNA called introns that are not used in the assembly of the protein. Furthermore, there are redundancies in the genetic code, so several different DNA sequences can correspond to the exact same sequence of amino acids. Many mutations within introns or between redundant sequences should therefore also be silent.

Since silent mutations do not affect how the organism interacts with its environment, both the probability that the mutation occurs and the probability that it gets passed on should not depend on where and how the creature lives. These mutations therefore are the ones most likely to accumulate at a steady rate and, by extension, to provide reasonable age estimates.

PATTERNS IN THE MUTATIONS OF HUMANS AND APES

Humans and other primates have provided a useful test case for determining whether the accumulation rate of silent mutations can be stable enough to serve as a timekeeper. Their DNA has been studied for many years, and recently the molecular biologists Feng-Chi Chen and Wen-Hsiung Li published a paper on this subject. They took DNA from a human, a chimpanzee, a gorilla, and an orangutan and obtained the sequences of fifty-three silent or noncoding regions (for a total of 24,234 base pairs from each animal). They then looked for a specific type of mutation known as a point substitution mutation. These occur when a single base pair is replaced with another base pair, for example when the sequence ACTG becomes ACCG. These sorts of changes are quantified in terms of the fraction of nucleotides that differ between the two sequences. For example, the sequences in Figure 5.8 have different nucleotides in five out of fifty positions, so the difference between these two sequences is 10%.[2]

With four animals, there are a total of six differences that can be measured: human-chimp, human-gorilla, human-orangutan, chimp-gorilla, chimp-orangutan, and gorilla-orangutan. Chen and Li calculated all of these differences (Table 5.1). From these six numbers, we can both deduce the relationships between these apes and argue that the mutation accumulation rate did not vary appreciably among the different animals.

Let us first reconstruct the relationships between these animals, assuming that they and their ancestors accumulated mutations at a steady rate. Relationships between animals are commonly depicted with a graph called a dendrogram or a phylogenetic tree, such as this:

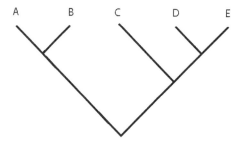

Table 5.1 Genetic differences between humans, chimps, gorillas, and orangutans. Measured by Chen and Li.

human-chimp	human-gorilla	human-orangutan
1.24%	1.62%	3.08%
	chimp-gorilla	chimp-orangutan
	1.63%	3.13%
		gorilla-orangutan
		3.09%

The letters at the top of the graph indicate a set of five animals living today, and the branching lines illustrate the ancestry of each of these organisms. In the recent past, all of them had distinctive ancestors, represented by the five separate lines leading to each letter. However, all of these creatures are ultimately derived from a single common ancestor, represented here by the point at the bottom of the plot. In between, the descendants of this common ancestor acquired mutations that set them apart from their relatives. In this case, the ancestors of creatures A and B diverged from the ancestors of creatures C, D, and E fairly early, while the ancestors of creatures D and E diverged from each other only recently.

We can begin to construct a phylogenetic tree for the great apes in Chen and Li's study by noting that the smallest of the six differences is between the human and the chimpanzee. This implies that these two animals have had the least amount of time to accumulate differences, so the ancestors of chimps and humans must have diverged after any split involving gorillas or orangutans. We can represent their relationship like this:

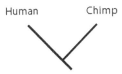

Next, we observe that both chimps and humans have fewer differences with gorillas (about 1.6%) than they do with orangutans (roughly 3.1%), so the divergence between the ancestors of chimpanzees and humans and the ancestors of gorillas occurred more recently than the split with the ancestors of orangutans. Humans, chimps, and gorillas therefore share a common ancestry illustrated by the following tree:

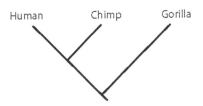

This leaves the orangutans, which are the most distinctive animals in this study. The ancestors of orangutans must have been accumulating distinct mutations for the longest period of time, so they are the first group to branch away from the common ancestor of all of these animals, as illustrated in the completed phylogenetic tree:

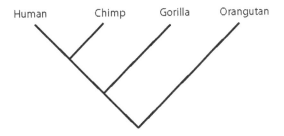

This graph encodes a brief history of the great apes, which starts with a species of primate that would ultimately give rise to humans, chimps, gorilla, and orangutans. At some point in the past, some descendants of this group began to acquire the mutations that would ultimately become the distinguishing traits of orangutans, such as extremely close-set eye sockets. Meanwhile, a different group of descendants acquired mutations that produced the characteristics seen in gorillas, chimpanzees, and humans. Later, this second group of animals itself broke into two groups, one that gives rise to gorillas and another that formed the ancestors of chimps and humans. Finally, the ancestors of humans and chimps diverge, and each acquires distinctive mutations. While it is likely that there were many other branching events similar to the three indicated in this graph, none of these events produced descendants that have survived to the present day.

The above diagram was generated assuming that all of these animals accumulated mutations at the same rate. We can check whether this was in fact the case by taking a closer look at the measured differences. If the ancestors of humans accumulated mutations faster than the ancestors of chimpanzees, then the measured difference between humans and gorillas should be larger than the difference between chimpanzees and gorillas, but this is not consistent with the data. The data instead indicate that the difference between humans and gorillas is almost identical to the difference between chimpanzees and gorillas, so the ancestors of humans and chimpanzees seem to have been accumulating mutations at nearly the same rate. Similarly, since gorillas, chimpanzees, and humans all differ from the orangutan by about 3.1%, the ancestors of all of these animals also appear to have acquired mutations at the same rate. Although this does not absolutely prove that the mutation accumulation rate was constant, it does tend to support the idea. Otherwise, it would be quite a coincidence if three different primates in three different environments all managed to accumulate an equivalent number of mutations over the same time period. More recent research has found that the mutation rates among the ancestors of the great apes did differ slightly, but the variations are sufficiently small (at most about 10%) that we can ignore them here.

Given that the great apes do appear to have accumulated mutations at an approximately constant rate, we can attempt to use these data to estimate when the various divisions actually occurred and, consequently, when our ancestors first acquired the ability to walk upright. We have already seen that since the number of differences between chimps and humans is smaller than the number of differences between orangutans and humans, the ancestors of chimps and humans must have parted company more recently than the ancestors of humans and orangutans. Now we will be more specific. The difference between chimps and humans (1.24%) is roughly two-fifths the difference between orangutans and humans (3.08%), so the chimp-human split must be that much more recent than the orangutan split. Integrating this additional information into the tree, we get this diagram:

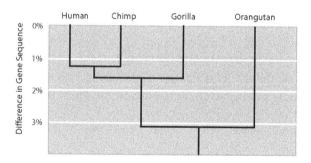

Here, the horizontal lines represent equal units of time, and the positions of the various branchings indicate when these events occurred. On the left are the percentage differences in the gene sequences. In this figure, the orangutan line splits off just before the time corresponding to 3% difference, and the human and chimpanzee lines split somewhat before the time that results in a 1% difference. Of course, we still need to figure out how many years it takes to accumulate a given variation. In the future, it may be possible to calculate the relevant timescales based on the molecular biology of the relevant organisms, but not yet. For now, mutation accumulation rates are estimated based on the fossil record.

While the fossil record of humans, chimps, and gorillas still does not provide enough information to indicate when exactly their ancestors diverged from the other apes, orangutans are another story. Paleontologists have found the fossils of an animal named *Sivapithecus*, whose skull has many features—such as close-set eye sockets and incisors of varying sizes—that among great apes are now seen only in orangutans. These similarities indicate that *Sivapithecus* is derived from the same line of animals that produced modern orangutans. These fossils have been found in deposits dating to around twelve million years ago, so the ancestors of orangutans must have started to acquire their distinctive traits sometime *earlier* than this.

Further information on the origin of orangutans has been gleaned from a fossil animal named *Proconsul*. This animal has features that are shared by all of the great apes—for example, it lacks a tail—but it has none of the features that are unique to chimps, humans, gorillas, or orangutans. Therefore, this creature probably existed before any of the living apes' ancestors acquired distinguishing characteristics. This beast is found in deposits from twenty million years ago, so ancestors of the orangutans probably acquired their distinctive characteristics sometime *after* this.

Together, the existence of *Sivapithecus* twelve million years ago and *Proconsul* twenty million years ago strongly suggests that the ancestors of the orangutans first diverged from the ancestors of the other apes about sixteen million years ago, give or take a few million years. Since it took sixteen million years for orangutans and other apes to acquire differences of 3.1% in these DNA sequences, the differences between two lines of great apes must increase at a rate of about 1% per five million years, allowing us at last to put a proper timescale on our family tree of the great apes:

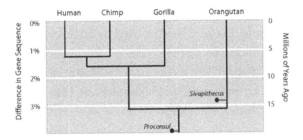

In addition, we can estimate that the lines leading to humans and chimps—which today differ by about 1.24%—began to diverge roughly 6.5 million years ago. This sort of analysis has been done many times using similar methods but different combinations of primates and fossil calibrators, and the results are usually the same.

Together, the molecular and the fossil evidence suggest that anthropologists are on the threshold of making a very exciting discovery. The fossils of *Australopithecus* demonstrate that our ancestors were already able to walk on two legs 4.5 million years ago. Meanwhile, the molecular data indicate that our ancestors began to acquire distinctively human-like traits—like bipedalism—only between six and seven million years ago. The first hominids to habitually walk on two legs therefore should have lived around five to six million years ago.

The newly discovered fossils of *Ardipithecus, Orrorin,* and *Sahelanthropus* date from this critical time period. Thus far, the available material is still too fragmentary to clearly document the earliest stages of hominid bipedalism, but as more and more remains of these and other similarly ancient hominids are found, they will hopefully document when, where, and how our ancestors began to walk upright. Such findings would be a boon to the study of human and primate evolution and also validate molecular methods of measuring the passage of time. Of course, more complete fossils of these animals could also surprise us and perhaps cause us to reevaluate our relationships with the other apes. In either case, the future for this field should be very interesting.

Beyond the study of human origins, new genetic sequence data are also having a major impact on many other areas of biology and paleontology. For instance, molecular data may provide important clues about the interrelationships and origins of modern groups of mammals like bats, rodents, primates, and whales. As we will see in the next chapter, this sort of research requires much more sophisticated analytical techniques to cope with the large amount of DNA sequence data involved as well as the comparatively large divergences between organisms. The reliability of these methods is somewhat uncertain at the moment, but they can still provide intriguing data and may eventually yield a clearer picture of mammalian evolution during the end of the age of dinosaurs.[3]

CLASSIFICATION[4]

CLASSIFICATION IS SORTING

… The basic idea behind biological classification is to sort organisms into similar, meaningful groups. Living and extinct species have a wide range of traits, yet share many common features. The challenge is to take into account their similarities and their differences in a way that reflects the basic biological foundations for their traits. Because the classification of all organisms follows the same general logic, if you learn the principles of how scientists classify organisms, you will learn much that can be applied to the classification of modern humans.

Like other aspects of biology, the scientific approach to biological classification has changed and developed over time. Although early classification systems were based on biological observations, they were not wholly scientific, did not reflect underlying biological mechanisms, and often they were not testable. Over the past 50 years these early classification approaches have been replaced by classification schemes that are more scientific, that rely on basic biological mechanisms, and are more testable. Most importantly, modern biological classification systems are grounded in knowledge about how species and populations have evolved and take into account their historical genetic relationships. Current biological classifications are hypotheses based on genetic relationships that can be supported or rejected by scientific studies.

Classification has an important place in the science of biology, but presents problems. Classification provides a shortcut to grasping the similarities and differences of organisms. If you know the classification of an organism you know quite a bit about it, even if you never have seen it. For example, if an animal is classified as an insect, you automatically know it will be rather small, have six legs, three body parts, and perhaps at least one pair of wings. It won't surprise you if the animal has a complicated, four-stage life cycle and breathes through a network of fine tubules that penetrate its tissues and cells. Modern classification systems also tell you about the relatedness of groups of organisms. One difficulty, though, is that each individual is unique. Not every kind of organism in a grouping will

---

3   Further reading: For information on human evolution at a popular level, try Ian Tattersal and Jeffrey Schwartz, *Extinct Humans* (Westview Press, 2000) and Carl Zimmer, *Smithsonian Intimate Guide to Human Origins* (Smithsonian Books, 2005). A good source on the web with many links is www.talkorigins.org/faqs/homs/. For more details, try Glenn C. Conroy, *Reconstructing Human Origins* (W. W. Norton, 1997).

For detailed information about the recently discovered ancient hominids, see the news article B. Wood, "Hominid Revelations from Chad," *Nature* 418 (2002): 133–136. Subsequent discussions of these early hominids can be found in K. Galik et al. "External and Internal Morphology of the BAR 1002'00 *Orrorin tugenensis* Femur" *Science* 305 (2004): 1450–1453, and T. D. White et al., "Asa Issie, Aramis, and the origin of *Australopithecus*," *Nature* 440 (2006): 883–889.

For potassium-argon dating, see R. E. Taylor and M. J. Aitken, *Chronometric Dating in Archaeology* (Plenum Press, 1997), chapter 4. A nice treatment can also be found in geology textbooks like Brian J. Skinner and Stephen C. Porter, *The Dynamic Earth*, 2nd ed. (John Wiley and Sons, 1992).

For the basics of genetics, a good place to start is Larry Gonick and Mark Wheeler, *The Cartoon Guide to Genetics* (Perennial Press, 1991).

Some books on reconstructing relationships from genetic data at the college level are Wen-Hsiung Li, *Molecular Evolution* (Sinauer, 1997) and M. Nei and S. Kumar, *Molecular Evolution and Phylogenetics* (Oxford University Press, 2000).

For the details of the genetic analysis cited in this chapter, see Feng-Chi Chen and Wen-Hsiung Li, "Genomic Divergences between Humans and Other Hominids," *American Journal of Human Genetics* 68 (2001): 444–456. For a more recent analysis of great ape DNA, see Navin Elango et al., "Variable molecular clocks in the hominoids," *Proceedings of the National Academy of Sciences* 103 (2006): 1370–1375.

4   Jan Jenner and Joelle Presson, "Biological Classification," *Biology: The Tapestry of Life*, pp. 405-413. Copyright © 2012 by Cognella, Inc. Reprinted with permission.

have the identical traits. Some species that are called insects, for example, have no legs. Furthermore, even within a grouping individuals are different. For instance, individual *E. coli* bacteria are not 100% identical. The challenge is to recognize individual variation yet to group the most similar organisms together. An opposite difficulty also must be considered: regardless of how different organisms may seem, all organisms share many features. For instance, elephants and fleas are more similar at a basic biological level than they are different. When constructing a classification scheme scientists must decide where to place their focus: on similarities or differences between organisms.

BIOLOGICAL CLASSIFICATION HAS A LONG HISTORY

People have been classifying organisms for as long as they have been naming them. The Western scientific classification of organisms has its roots in the study of nature by scholars in ancient Greece. Interest in classification flourished in Europe in the 1600s to 1900s, and continues as a major scientific field of study today. Over this time period two approaches to classification can be identified. Before the mid-1900s most approaches to classification relied on physical traits and attempted to classify individuals into distinct and separate kinds of life. This trait-based, distinct approach to classification is often called **phenotypic classification**. Phenotypic classification dates to the earliest work of Greek scholars, and this approach dominated the work of the European naturalists from the 1600s to the early 1900s. In the mid-1900s a few biologists introduced the alternate idea of classification by evolutionary/genetic relationships, or **phylogenetic classification**. Phylogenetic classification emerged once the principles of evolution became firmly established as the best scientific explanation for biological diversity. It rests on the understanding that all living things are genetically related, and that the degree of relationship between different groups of organisms can be understood. Although phylogenetic classification is predominant today, aspects of phenotypic classification are still useful, so it is important that you understand both philosophies of classification. For one thing, many aspects of phenotypic classification nomenclature are embedded in the language and structure of classification systems. In addition, while the usual concepts of human races are not supported by scientific evidence, they are rooted in the phenotypic classifications of the early European naturalists.

PHENOTYPIC CLASSIFICATION CONSIDERS PHYSICAL SIMILARITIES

The term *phenotypic classification* gets its name from the term **phenotype**, which refers to the observable traits of an individual. Phenotype is contrasted with **genotype**, which refers to the genes that determine those traits. Phenotypic classifications rely on such features as the size, shape, and structure of organisms. An organized phenotypic classification system first arose in the 1700s, and was formalized by Carolus Linnaeus. These early classification efforts focused on **species**, the basic kinds of biological organisms. Linnaeus transformed the haphazard and inconsistent classification of his time, replacing it with a naming system that used the similarities and differences of their traits to group species in a hierarchical way. In the Linnaean system the species is the basic unit of biological classification. Populations are placed into the same species when their traits are similar, and are placed in different species when their traits are different. For example, humans share phenotypic traits such as similar anatomy, upright posture, large brains, grasping hands, and stereoscopic vision. On this basis, Linnaeus grouped all modern humans into a single species. Because of phenotypic differences, humans and chimpanzees are grouped into different species, despite their many similarities. In the Linnaean hierarchy similar species are placed into a larger group called a **genus**, and different genera have different traits. Using modern humans again as an example, our species is given the unique scientific name, *Homo sapiens*. This two-word name incorporates the name of our genus, *Homo*, plus a distinctive word to identify the species. In the case of humans, the species word is *sapiens*. So, in the Linnaean system modern humans are *Homo sapiens*, Latin for "wise man." Despite the move away from

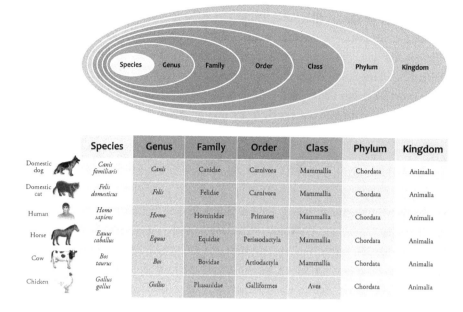

	Species	**Genus**	**Family**	**Order**	**Class**	**Phylum**	**Kingdom**
Domestic dog	*Canis familiaris*	*Canis*	Canidae	Carnivora	Mammalia	Chordata	Animalia
Domestic cat	*Felis domesticus*	*Felis*	Felidae	Carnivora	Mammalia	Chordata	Animalia
Human	*Homo sapiens*	*Homo*	Hominidae	Primates	Mammalia	Chordata	Animalia
Horse	*Equus caballus*	*Equus*	Equidae	Perissodactyla	Mammalia	Chordata	Animalia
Cow	*Bos taurus*	*Bos*	Bovidae	Artiodactyla	Mammalia	Chordata	Animalia
Chicken	*Gallus gallus*	*Gallus*	Phasanidae	Galliformes	Aves	Chordata	Animalia

Figure 5.9: The Linnaean classification system. Major categories in this hierarchical classification system are, from smallest to most inclusive, species, genus, family, order, class, phylum, and kingdom. Notice that although all these animals are grouped into the same kingdom and phylum, they belong to two different classes. Notice that the mammals belong to five different orders, many different families, genera, and species. Notice also that each species has a two-word scientific name that identifies its genus and species.

Linnaean classification in the 20th and 21st centuries, scientists today still use the genus-species naming system. Linnaeus gathered similar genera into a larger group called a family; similar families are grouped into an order; similar orders are grouped into a class; similar classes are grouped into a phylum; and phyla are grouped into a kingdom (Figure 5.9).

At the highest level of Linnaean classification, kingdoms are groups that contain organisms that share profound common characteristics. For instance, the Plant Kingdom contains multicellular photosynthetic organisms, while the Animal Kingdom contains multicellular organisms that are non-photosynthetic and obtain food by eating other organisms. Over the years biologists debated how many kingdoms to define. Some suggested five kingdoms, others said eight, and still others said thirteen. Most schemes name five kingdoms: Bacteria (Monera), Protists, Fungi, Plants, and Animals (Figure 5.10). Even though these five kingdoms are used today, the five-kingdom system does not reliably reflect the genetic relationships of groups of organisms. You will explore this idea later in this chapter.

Animals   Plants   Fungi   Protists   Monera

Figure 5.10: An example of the Five Kingdoms based on Linnaean classification principles.

In its time the Linnaean classification scheme was a significant advance. It encouraged researchers to pay attention to detailed observations about the traits of organisms. Linnaean classification also pushed researchers to examine similarities and differences between species. Still, the Linnaean classification system had its problems. The criteria for classification were not clear and explicit; often they were based on the opinions of individual scientists. In many cases there was no clear way to scientifically test which classification scheme was more likely to be correct. For example, should algae, which are photosynthetic organisms, be placed in the same phylum or kingdom as plants, or not? When it was discovered that *Euglena*, a unicellular organism found in freshwater, had both animal and plant characteristics, it was difficult to classify them. Most importantly, the Linnaean idea that species are distinct and separate kinds was not supported by subsequent scientific work. Certainly Linnaeus and other naturalists of his era recognized that individuals vary along many traits, but nevertheless they saw fixed boundaries between one species and another. For example, in the Linnaean system there is no way to explore whether humans are more closely related to chimpanzees or to gorillas. Each is simply a different species. The phylogenetic approach to classification eliminated many of these difficulties.

The most familiar ideas about race—those that define people as Whites, Blacks, Hispanics, etc.—have their roots in the Linnaean approach to classification. While these racial categories do have strong cultural connotations, often there is an underlying assumption that races are "distinct and separate biological kinds" of people. This idea is quite similar to the Linnaean idea of species, genera, and other categories as being "distinct and separate kinds" of organisms. The evidence from the last century of biological research causes scientists to reject this Linnaean approach, both to biological classification in general and to classification of humans in particular. If you look at human diversity from the perspective of genetic and evolutionary evidence, you arrive at an approach to classification that looks at how closely people are related.

PHYLOGENETIC CLASSIFICATION IS BASED ON GENETIC RELATIONSHIPS

Today biological classification has progressed and morphed into a more precise science. A major advance came from the work of Charles Darwin. He articulated the principles of evolution as the mechanism for how biological diversity emerges and more than 250 years of subsequent scientific studies have shown that Darwin's principles of evolution are correct. Building on Darwin's work, genetics research has shown that the principles of evolution are explained by the structure and function of DNA. These two lines of thought, evolutionary biology and DNA science, have been integrated into one phylogenetic understanding of diversity. Now there is a way to classify organisms that reflects their genetic and evolutionary history.

Phylogenetic systems often use physical traits to classify organisms, but they focus on traits that reveal genetic ancestry. To put this in human terms, consider a project to classify all of the individuals in a village in central Asia. A phenotypic classification would place people with dark hair, blue eyes, tall height, and long noses into one group, while people with light hair, dark eyes, short height, and short noses might be placed into another. In contrast, a phylogenetic classification would focus on family history. In essence, a phylogenetic classification produces a family tree. This approach is less direct and can be a bit more difficult to carry out. For example, in a village in central Asia, scientists might first try to establish family histories, but people can be wrong about the details of their ancestry. So, in addition to verbal evidence, scientists would seek objective measures of genetic relationships, such as public records and unusual inherited traits. Analyses of DNA would be even more important. DNA is the molecule that is passed on from one generation to the next. It carries the information that determines the traits of an individual. The result of a phylogenetic study would be a documented family tree showing who was related to whom by common ancestors. In the central Asian village population you might have the Han family, the Le family, and the Phan family. The individuals in one family tree might have similar traits, such as a tendency toward certain diseases, but their distinct genetic relationships would be the basis for

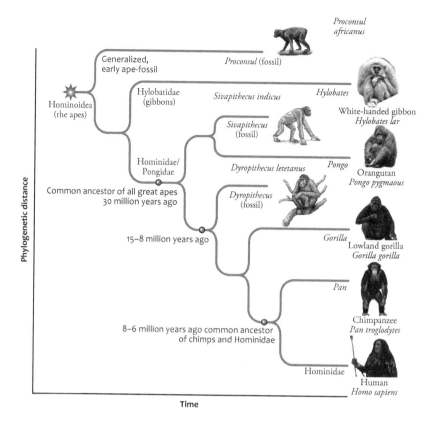

Figure 5.11: The phylogenetic tree of living and some extinct apes. The blue star indicates the common ancestor of all apes.

the classification. More importantly, the individuals in each family might have complex genetic relationships, and the groups might overlap. There are not likely to be strict and distinct boundaries between the families. This is the essence of phylogenetic classification.

Phylogenetic analyses help scientists to construct **phylogenetic trees**. These are family trees of phyla, species, populations, or any other biological groups you care to identify and study. A phylogenetic tree shows the genetic relationships between these groups. For example, Figure 5.11 shows a phylogenetic tree for some living and extinct apes. The tree is based on phenotypic traits and on DNA analyses. While there are often no axes drawn on a phylogenetic tree, two axes are implied on all phylogenetic trees: time and genetic difference. In Figure 5.11 the x-axis (horizontal) is time, with earlier times to the reader's left. The y-axis (vertical) is genetic or phylogenetic difference. As you can see, humans and chimpanzees shared a common ancestor about 6 to 8 million years ago (x-axis), and are separated by less genetic distance than humans and gibbons (y-axis). Each branch point on the tree represents an unnamed common ancestor of all the species beyond that branch point. For example, the star indicates the common ancestor of all known apes, including humans. Scientists can be fairly confident of when the common ancestor of later groups lived. Based on your reading of this figure about how long ago did all living apes, including gibbons, share a common ancestor?

Another point made by Figure 5.11 is that not all species that emerge through natural selection have survived to the present time. The fossil record indicates that around 5 to 20 million years ago many species of apes ranged from Africa through Asia and Europe. Today only five major species of apes survive. The rest went extinct. The living species can be studied using DNA and other traits. Although DNA analysis of some fossil species has been successfully accomplished, this is highly unusual. Extinct species and their phylogenetic relationships usually can only be known from fossils.

CELLS ARE CLASSIFIED INTO THREE PHYLOGENETIC GROUPS

Thus far you might have the impression that phenotypic and phylogenetic classification systems are completely unrelated, and result in different groupings of organisms. Phenotypic classification is based on physical traits, while phylogenetic classification is based on genetic lineage and evolutionary history. These classification systems, however, often have similar results. How can this be possible? The answer lies in the fact that phenotypic traits are the result of genetic and evolutionary processes and mechanisms. Individuals with similar phenotypic traits can have very different genetic and evolutionary histories—different family trees—but they often have similar evolutionary histories. Although the results of a phenotypic classification system may be similar to those of a phylogenetic classification, sometimes they are quite different. Unlike the phenotypic system, the phylogenetic classification system will provide a clear way to test if it is likely to be correct, and it will reveal the genetic relationships between populations.

You can see an example of the overlap between phenotypic and phylogenetic analyses in the classification of cell types. In the traditional phenotypic scheme, *prokaryote cells* are defined by their lack of complex internal structures. In contrast, *eukaryote cells* are defined by their complex internal structures such as a nucleus and other organelles. The terms prokaryote and **eukaryote** are still in common use because they describe basic structural variation of cell types.

Analyses of DNA show more subtle phylogenetic patterns and reveal the history of genetic and evolutionary relationships among cells. These studies show that the phenotypic *Eukaryotes* are indeed a large, genetically related group. Thus far the two kinds of classification systems have similar results. A genotypic analysis of the cells called prokaryotes, however, reveals they are actually two genetic groups of cells, one called *Bacteria* and the other called *Archaea*. Although they look similar, Bacteria and **Archaea** differ genetically and biochemically. Figure 5.12

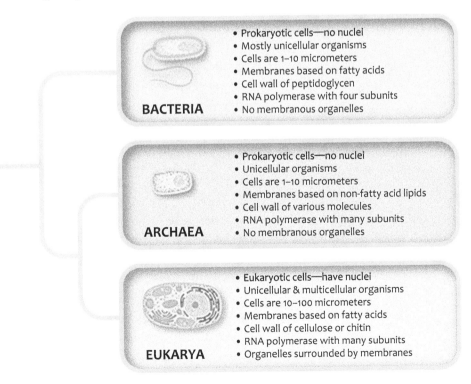

BACTERIA
- Prokaryotic cells—no nuclei
- Mostly unicellular organisms
- Cells are 1–10 micrometers
- Membranes based on fatty acids
- Cell wall of peptidoglycen
- RNA polymerase with four subunits
- No membranous organelles

ARCHAEA
- Prokaryotic cells—no nuclei
- Unicellular organisms
- Cells are 1–10 micrometers
- Membranes based on non-fatty acid lipids
- Cell wall of various molecules
- RNA polymerase with many subunits
- No membranous organelles

EUKARYA
- Eukaryotic cells—have nuclei
- Unicellular & multicellular organisms
- Cells are 10–100 micrometers
- Membranes based on fatty acids
- Cell wall of cellulose or chitin
- RNA polymerase with many subunits
- Organelles surrounded by membranes

Figure 5.12: Phylogenetic classification reveals three fundamental cell types. Notice that the two groups of prokaryotes are distinguished biochemically, not phenotypically. Notice also that archaeans are more closely related to eukaryotes. What characteristic(s) do they share?

shows the fundamental family tree of all cell types. In this system the three fundamental cellular groups are called *Domains*. The branching pattern reflects genetic relationships within each Domain. Notice that the Eukaryotes, including yeasts, sponges, plants, fungi, animals, and many more, are more closely related to the Archaea than they are to the Bacteria.

There is one other important point about the phylogenetic classification of cells shown in Figure 5.12. Notice that all three cell types are joined by a single line that goes back to the common ancestor of all cells. Unlike the kingdoms of the Linnaean system, the phylogenetic classification emphasizes that all organisms, whether living, dead, or extinct, are related because they all are descendants of a common ancestor. There are differences between the three groups, but they are more similar than they are different. All three groups share a great deal of DNA because they all inherited it from a common ancestor, just as you and your cousins inherited common DNA from your great-great-great-grandmother.

A PHYLOGENETIC CLASSIFICATION OF EUKARYOTES

The phylogenetic classification approach can be applied to any set of organisms. The classification of eukaryotes is an example. In the typical phenotypic five-kingdom classification scheme eukaryotes are divided—based on phenotypic traits—into Protists, Fungi, Animals, and Plants. The place of algae in this scheme has been open to debate. Some researchers have grouped algae with the protists, while others have placed them with the plants. Figure 5.13 shows the most recent phylogenetic scheme. This way of depicting the relationships is a bit different from what you probably are used to. In this figure the common ancestor of eukaryotes is at the center of the irregular star. The phylogenetic distance is the distance from the center of the star *and* the distance around the circumference of the star. This phylogenetic analysis shows that Animals, Plants, and Fungi emerge as major phylogenetic groups. Notice that the fungi are actually more closely related to animals than they are to plants. The protists and algae, however, sort very differently in this scheme than they do in phenotypic schemes (see Figure 5.11). Some, such as the brown algae, are in groups by themselves. Others share similar DNA and thus genetic lineages with major multicellular eukaryote groups. You can see this more clearly in Figure 5.14 which shows the portion of the eukaryotic phylogenetic tree for plants, fungi, and animals. Here the protists called choanoflagellates are revealed as most closely related to eukaryotic animals. The green algae, also called protists in phenotypic classification schemes, are related to modern plants. The lesson from phylogenetic studies is that around two billion years ago natural selection allowed a dramatic diversification of single-celled eukaryotes. Some of these gave rise to the plants, fungi, and animal groups known today, while other single-cell eukaryotes led to unique phylogenetic groups. This understanding never would have emerged from a Linnaean classification approach that focused only on traits, rather than on phylogeny.

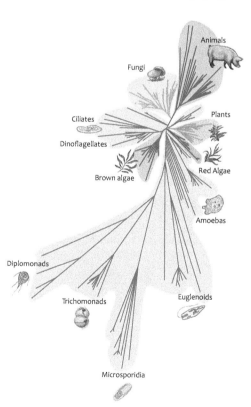

Figure 5.13 Phylogenetic groups of eukaryotes. Biochemical investigations reveal great diversity among groups of protists. Notice that fungi and animals are closely related.

PHYLOGENETIC TREES WITHIN SPECIES

Just as phylogenetic trees can be made of different species, genera, and kingdoms, a phylogeny of human populations can be constructed that reveals their relationships. ... For now, you can gain an intuitive understanding of the phylogenetic trees within species by considering your own family tree. Figure 5.15 shows the framework of a family tree. Fill in what you know of your own family history, expanding the tree as far back as you can. ...

For instance, you have two parents, four grandparents, eight great grandparents, etc. And this is only three generations—probably about 150 years—of ancestors. How many ancestors would you have if you went back 50,000 years? Thousands. This

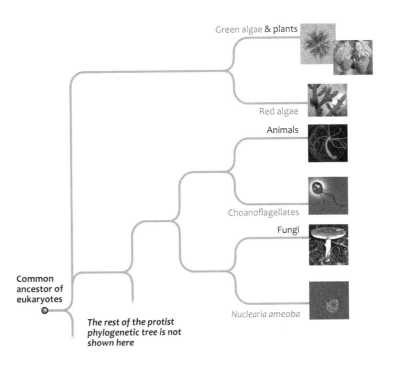

Figure 5.14: Partial phylogenetic tree of some organisms called "protists." Organisms that once were grouped as protists are shown in red.

is something of a paradox. Clearly there were fewer humans alive 50,000 years ago than there are today. How can each person have more and more ancestors further back in time? The answer, of course, is that you *share* ancestors with other people alive today. As you go back farther in time you share ancestors with more and more living people. Can you see where this logic takes you? There is a point in human history when all humans share a pair of common ancestors: one male and one female (Figure 5.16). Just as the principles of phylogenetic classification predict that all organisms share a common ancestor that lived nearly four billion years ago, a scientific approach to human diversity predicts that all humans share a single pair of common ancestors, one male and one female.

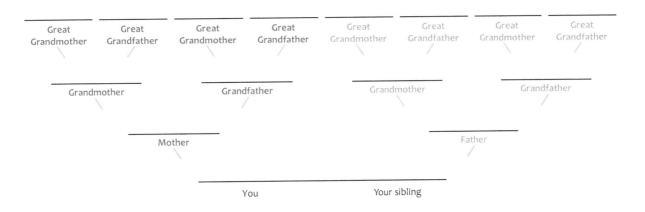

Figure 5.15: Constructing your family tree. Use this figure to fill in your own family tree. Do you know more than the relatives listed here? Who are your cousins, aunts, and uncles? How do they fit into this tree? Notice how the number of your relatives increases as you go back in time.

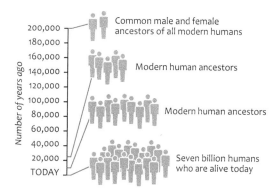

Figure 5.16: Common ancestors. All humans share a set of common ancestors who first emerged about 200,000 years ago.

BUILDING A PHYLOGENETIC TREE

The idea of phylogenetic classification still may seem quite abstract to you. It will strengthen your understanding of the idea if you work through an example that demonstrates the principles of phylogenetic classification. The aim of this example is to illustrate the phylogenetic relationships of four groups of organisms, which are actually four different groups of primates. You think they are closely related, but you want to know the overall shape of the family tree. Which groups and traits emerged first; which came next; which were last? Which traits are common to all groups and so are older traits, and which have evolved more recently? This process makes several assumptions, all of which are well-supported by research. The first assumption is that any population has some traits that are inherited intact from distant ancestors, and other traits that are more recently evolved. Second, evolved traits emerged in a temporal order, some before the others. Third, the pattern of traits within and between living populations can be used to construct this evolutionary history.

One important concern is to figure out which traits are inherited from a more distant ancestor and which traits are more recently evolved. The ancestral traits will be shared by all groups under study and can be used to hypothesize the traits of the ancestral populations. Traits that have evolved more recently will be used to reconstruct the evolutionary history of the four groups of primates. Biologists have special terms to refer to these two kinds of traits. **Shared ancestral traits** are those that a population inherits from, and so shares with, a more distant ancestor. **Shared derived traits** are those that have more recently evolved, and so are expressed exclusively by the populations you wish to classify. Here is an example. Humans are warm-blooded, gestate their young in their bodies before birth, and have four appendages. Chimps have these same traits. These are ancestral traits shared by all mammals. Chimps and humans both have large brains, frontally placed eyes, and opposable thumbs—all shared derived traits. These shared derived traits suggest that humans and chimps share a recent common ancestor. How can you figure out which traits are shared ancestral and which are shared derived? Often you do not have to look into the past—such as the fossil record—to figure this out. Ancestral traits are present in populations that are alive today. In other words, some living groups have not changed much—at least in certain characteristics—from their long-lost ancestors. For example, modern bony fish have many newly evolved traits, such as distinct colors and complex behavior patterns. On the other hand, modern bony fish also have many ancestral traits, such as their lack of physiological regulation of body temperature and lack of limb bones. Any study of the phylogenetic relationships amongst populations must include at least one group that is likely to provide the ancestral character traits. A comparison group that is included to help reveal the ancestral traits is called an **outgroup**.

LINEAGES AND CLADES[5]

… Several terms pertain to the study of species in the past (Figure 5.17). A *lineage* is single line of descent. One speaks of the human lineage and the reptilian lineage. **Extinction** is the termination of a lineage and marks the end of the line—a lineage that failed to survive to the present. Extinction is the eventual fate of all species, as

5   Richard J. Huggett, "Biogeographical Processes I: Speciation, Diversification, and Extinction," *Fundamentals of Biogeography*, pp. 25–26. Copyright © 2004 by Taylor & Francis Group. Reprinted with permission.

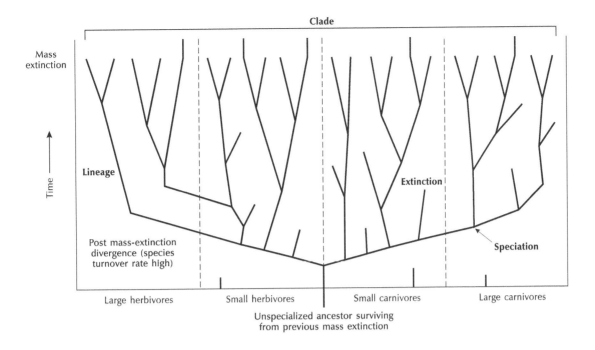

Figure 5.17: Lineage, speciation, extinction, and clade, explained.

discussed later in this chapter. In the fossil record, *speciation* is the branching of lineages. In other words, it marks the point where a single line of descent splits into two lines that diverge from their common ancestor. It occurs when a part of a population becomes reproductively isolated from the remaining populations of the established species by any of the mechanisms mentioned in the previous section. It is well nigh impossible to reconstruct the isolating mechanisms involved in fossil populations, although evidence may sometimes exist for geographical isolation.

Fossil assemblages traced through time reveal clades. A **clade** is a cluster of lineages produced by repeated branching (speciation) from a single lineage. The branching process that generates clades is *cladogenesis*. The clade Elephantinae comprises two extinct genera—*Primelephas* and *Mammuthus*—and two living genera—*Elephas* (modern Asian elephants) and *Loxodonta* (modern African elephants). Evidence from mtDNA suggests that *Elephas*, *Loxondonta*, and *Mammuthus* started to differentiate in the Late Miocene epoch around 5.6–7.0 million years ago. This supports fossil finds since 1980: the oldest known *Loxodonta* specimen comes from Baringo, Kenya, and Nkondo-Kaiso, Uganda, between 7.3 and 5.4 million years ago; and the oldest known *Elephas* comes from Lothagam, Kenya, about 6–7 million years ago (Tassy and Debruyne 2001). ...

TYPES OF PHYLETIC GROUPINGS

We use the term **monophyletic** to describe a *true* clade, a group of organisms and their most recent common ancestor. A **paraphyletic** group of organisms is similar to a monophyletic group; it also includes the most recent common ancestor but excludes certain lineages of descendants. A **polyphyletic** group is composed of organisms that share similar characteristics, such as marine mammals, but do not share a direct common ancestor.

SPECIATION, NATURAL SELECTION, AND TYPES OF EVOLUTION[6]

... In many cases the adaptation of a species to its environment seems perfect, but closer examination shows variation among individuals within the population. One example of what seems to be a near perfect fit produced by natural selection is the golden-winged sunbird of East Africa. This delicate bird has a long, curved beak that seems an exact fit for the tubular mint flowers from which it gathers nectar (Figure 5.18A). The match of its bill to the curve of the mint flowers is so close that a typical golden-winged sunbird is able to gather 90% of the nectar in a mint flower in just 1.3 seconds, a record unmatched by similar species whose bills are not so well adapted to the shape of mint flowers (Figure 5.18B). Nevertheless, there is variation in bill size and shape of golden-winged sunbirds. Some have bills that are slightly longer; some are slightly shorter, some are slightly more curved, and so on. This variation in bill shape means that some of the golden-winged sunbirds are more efficient at gathering nectar than are others.

Bill-shape variation provides the possibility for the population to change in response to environmental changes. Consider what would happen if part of the region where sunbirds live experienced a long period of colder and longer winters. A shorter growing season would mean that mint flowers would be smaller. Under these conditions birds with smaller bills would be more efficient at gathering nectar, would survive in greater numbers, and would

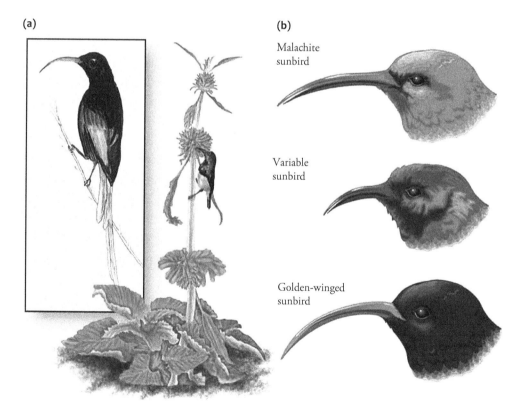

(a)

(b)

Malachite sunbird

Variable sunbird

Golden-winged sunbird

Figure 5.18: A close match of bill shape and flower shape. (a) The golden-winged sunbird can get nectar from mint flowers faster than two other sunbird species (b).

---

6   Jan Jenner and Joelle Presson, "Life Evolves: Darwin and the Science of Evolution," *Biology: The Tapestry of Life*, pp. 352, 354-355, 356-359. Copyright © 2014 by Cognella, Inc. Reprinted with permission.

have more offspring than birds with larger bills. Over several generations the typical bill for a golden-winged sunbird in this region would begin to change—and overall the population would have smaller bills than they did when the seasons were warmer. Of course, this can happen only if some of the birds in the original population had bills adapted to feed on smaller flowers. If individuals could not find flowers suited to their bill shape, the golden-winged sunbirds would die out and might become locally extinct.

MANY STUDIES TRACE NATURAL SELECTION IN ACTION

Because the individuals of some species live such long lives in comparison with human observers, in many cases it is difficult to see evolution happening. Other species have short life spans and researchers can observe changes in a population over generations. [...]

House Sparrows provide a good example. House Sparrows are native to Eurasia, but in the 1850s several were released in Brooklyn, New York. By 1900 this hardy, adaptive, aggressive species had invaded suitable habitats around cities and towns all across North America and even had reached Vancouver, British Columbia, on the Pacific Coast of Canada. By 1914 House Sparrows had reached Mexico City, and in 1974 they were found in Costa Rica. Figure 5.19 shows House Sparrows from different locations. Notice that the populations of House Sparrows in different regions are subtly different, even though they are the same species. In general, northern populations are larger and darker than are southern populations. These differences are evidence of natural selection in action. As House Sparrows spread out from New York City, they encountered different environments that favored the survival of House Sparrows with slightly different traits.

A more thorough understanding of natural selection comes from studies of species of finches on the Galápagos Islands, collectively called Darwin's finches. One feature that distinguishes one species of Darwin's finches from another is the size and strength of the birds' beaks (Figure 5.20). The size and shape of a finch's beak is an inherited trait, and it is related to the type of food that each finch species eats. Species with larger, heavier beaks can break open and eat harder seeds, while finch species with smaller beaks are restricted to feeding on smaller, softer

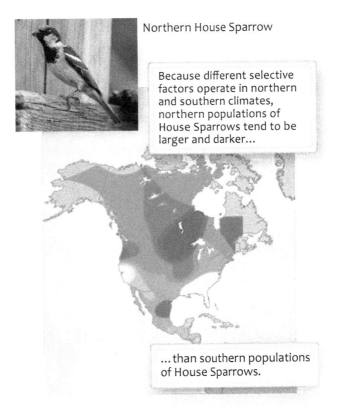

Northern House Sparrow

Because different selective factors operate in northern and southern climates, northern populations of House Sparrows tend to be larger and darker...

...than southern populations of House Sparrows.

Figure 5.19: Geographic changes in House Sparrows. The geographic variation in size and color of House Sparrows in North America is evidence for natural selection. Large size, indicated by darker orange on the map, is an adaptation to cold; northern House Sparrows are larger than southern House Sparrows. Northern House Sparrows also are darker than southern House Sparrows.

1. Medium ground finch
Geospiza fortis

2. Large ground finch
G. magnirostris

3. Sharp-beaked ground finch
G. difficilis

4. Small ground finch
G. fuliginosa

5. Large cactus finch
G. omnirostris

6. Cactus finch
G. scandens

Figure 5.20: A comparison of beaks of some species of Darwin's finches. Different species of Darwin's finches have beaks that vary in size and shape. Beaks are adaptations to feeding on specific kinds of foods. Larger beaks can crack harder seeds; smaller beaks can manipulate and open smaller, softer seeds.

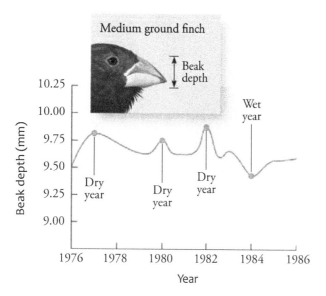

Figure 5.21: During dry years Galapagos medium ground finches have larger beaks. In dry years seeds are larger, harder, and fewer. In wet years seeds are smaller, softer, and more abundant.

seeds. Of course, within any species, beak size and strength vary between individuals.

During most seasons there is enough rain to produce seeds of varying hardness on the Galápagos Islands, and so birds with a variety of beaks are able to survive. During droughts, however, seeds dry out and on average they become harder and more difficult to crack open. So birds with larger beaks are able to eat more seeds than are birds with smaller beaks. Of course, the birds that eat more are more likely to survive and to have offspring than birds with smaller beaks that get less to eat. The principles of natural selection lead to the hypothesis that after several seasons of drought the percentage of individuals in the population with slightly larger bills will increase.

Measurements of beak sizes from 1977 to 1984 provide direct support for this hypothesis. These were years of severe drought in the Galápagos Islands, and the dried-out seeds were hard and difficult to break open. Figure 5.21 shows that at the end of two seasons of drought, the percentage of individuals in the population with larger beaks increased. And, because the size of the beak is determined by the **alleles** a bird carries, this change in beak size over generations reflects a change in the frequencies of alleles for larger or smaller bills within each population.

Eventually, the environment returned to its normal, moister state. Seeds became softer, and smaller seeds became more abundant. Accordingly, finches with smaller, weaker beaks could survive and reproduce in greater numbers. Think, though, about what would have happened if drought had become an environmental constant: seeds would have stayed hard, and each original finch species that survived would have evolved into a species with a slightly larger-sized beak. …

When Darwin faced the problem of explaining how new species emerge, he had insights that were grounded in his experiences with English country life. Darwin knew that farmers could change the features of their domestic

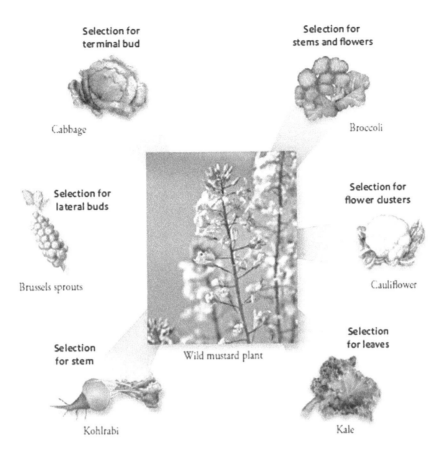

Selection for
terminal bud

Cabbage

Selection for
stems and flowers

Broccoli

Selection for
lateral buds

Brussels sprouts

Selection for
flower clusters

Cauliflower

Selection
for stem

Wild mustard plant

Selection
for leaves

Kohlrabi

Kale

Figure 5.22: Artificial selection in mustard plants. By selectively breeding plants with particular traits, all of these domestic vegetables have been bred from wild mustard plants. (Andreas Trepte)

animals and crop plants by carefully controlling breeding. Controlled breeding that produces new strains of domestic organisms is called **artificial selection**. For example, dogs have been bred to produce canines as different as Chihuahuas and Neapolitan mastiffs; cats have been bred to produce the furless sphynx as well as fluffy Persians. As a result of selective breeding, every season brings new varieties of flowers to garden centers and different varieties of fruits and vegetables to supermarkets. Selective breeding was no less amazing in Darwin's time and in it he saw evidence that whether artificial or natural, selection can dramatically change the average traits of a population. Figure 5.22 shows the results of one of the best modern examples of artificial selection. All of the vegetables illustrated here originated from the selective breeding of wild mustard plants.

THREE TYPES OF NATURAL SELECTION HAVE BEEN IDENTIFIED

Natural selection can produce three different outcomes. First, natural selection could push the average value of a trait in one direction. Second, natural selection could keep the average value of a trait the same but eliminate extreme values of the trait. And finally, natural selection could produce two different groups within a population, each with its own average value of a trait. Before considering examples of each of these, it will help to consider what a graph of the values of a trait in a population would look like.

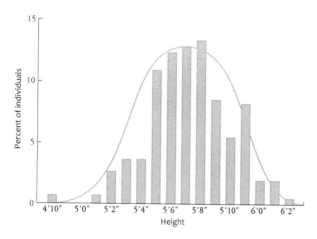

Figure 5.23: Distribution of height among men. There were only 200 men in this sample. A larger sample might have more individuals at either end of the curve.

Figure 5.23 shows the distribution of height among men. The average value of height is about 5 feet 6 inches (1.6 meters), but some men are well under 5 feet tall and others are well over 6 feet tall. A graph shaped like this, in which most values cluster around the average and fewer values are at the extreme, is called a *normal distribution*. The effects of the different types of natural selection can be evaluated by changes in a normal distribution that take place over generations.

First, consider what happens if natural selection changes the average value of a trait. This type of natural selection is called **directional selection**. One striking example is happening right now in species of ocean fishes caught by fishing fleets. For many years humans have taken the larger fishes and have thrown smaller individuals back into the sea. As a result, the average size of the fishes that are caught has been decreasing. Some studies suggest that these fishing practices are changing the genetic traits of populations of fishes (Figure 5.24). Researchers maintained three tanks of fishes over four generations. Each generation 90% of the fishes in each tank were removed, but the fishes that were removed differed. In one tank a random 90% was removed; in the second tank the largest 90% were removed; in the third tank the smallest 90% were removed. After four generations of this practice, the fishes in the first tank grew to be about the same size as they had been in the first generation. The fishes in the second tank grew to be much smaller than they had been in the first generation. The fishes in the third tank grew to be much larger than they had been in the first generation. This demonstrates that by removing fishes with the alleles that allow growth to a large size, modern fishing techniques are unintentionally pushing the populations of ocean fishes toward a smaller size.

Stabilizing selection occurs when individuals with extreme values of a trait do not survive to breed and so the population clusters about the average values for the traits. One striking example is birth weight in humans (Figure 5.25). Babies that weigh less than 5 pounds or over 9 pounds are much more likely to die during birth than are babies that weigh between these values. This differential mortality leads to a narrow distribution of live birth weights. Since the late twentieth century medical advances have allowed higher rates of survival in both small and large babies. If this trend continues, in 25 or 50 years the birth weight distribution might look broader, which will mean that modern culture is changing human evolution.

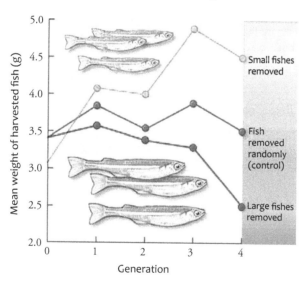

Figure 5.24: An experiment to determine the effect of fishing on size of fishes. Small fishes were removed from one population of captive fishes. Large fishes were removed from a second population. Fishes were randomly removed from the control population. After four generations of removing fishes, the average weight of fishes in the large and small populations had changed, while the control group was unaffected. This experiment shows that fishing practices can change the inherited traits of fish populations.

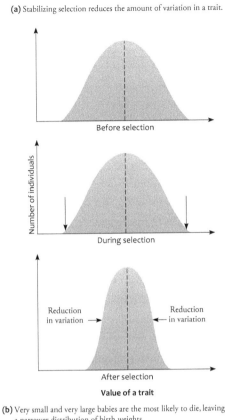

(a) Stabilizing selection reduces the amount of variation in a trait.

Before selection

Number of individuals

During selection

Reduction in variation → ← Reduction in variation

After selection

Value of a trait

(b) Very small and very large babies are the most likely to die, leaving a narrower distribution of birth weights.

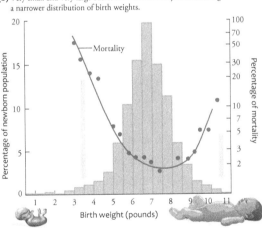

Mortality

Percentage of newborn population

Percentage of mortality

Birth weight (pounds)

Figure 5.25: Stabilizing selection is selection for the mean.

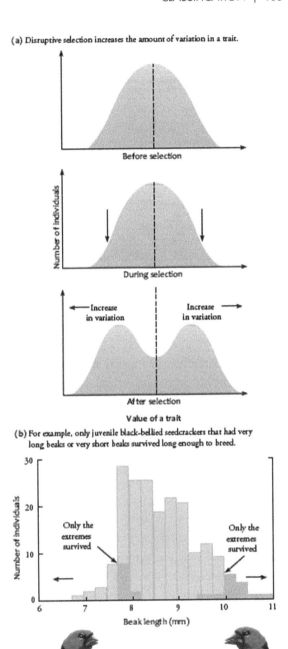

(a) Disruptive selection increases the amount of variation in a trait.

Before selection

Number of individuals

During selection

←—Increase in variation   Increase—→ in variation

After selection

Value of a trait

(b) For example, only juvenile black-bellied seedcrackers that had very long beaks or very short beaks survived long enough to breed.

Number of individuals

Only the extremes survived

Only the extremes survived

Beak length (mm)

Figure 5.26: Disruptive selection.

Finally, **disruptive selection** pushes a population away from the average value of a trait, leading to a distribution that has two average values, or a *bimodal distribution*. Figure 5.26 shows an intriguing example among black-bellied seed-cracker birds in West Africa. The adults of this population have beaks that are either long or short, but usually not of intermediate size. The reason seems to be that only two sizes of seeds are available—and birds tend to specialize, taking large seeds or small seeds. Birds with intermediate-sized beaks tend to die before they reach adulthood. If this pattern continues into future generations, it is possible that the small-beaked and large-beaked black-bellied seed-crackers will split into two different species.

Figure 5.27: Sexual selection. The gorgeous, but impractical tail feathers of a male peacock are the result of sexual selection.

SEXUAL SELECTION IS A SPECIAL CASE OF NATURAL SELECTION

Among animal species selection by the opposite sex is one aspect of the environment that determines which individuals mate and have offspring. Darwin was one of the first to suggest that mate choice could drive evolution. Females are especially choosy about which males they will mate with, and these choices can have a large impact on the traits of future generations, especially the males of the population. The peacock's feathers, the bright plumage of many male birds, and the bright colors of many male fishes are traits that may be driven by female mate choice (Figure 5.27). Males without these traits are less likely to find mates and do not breed. Why do females prefer showy males? The best hypothesis is that showy males are on average stronger and healthier than are drab or small males. Therefore, females who mate with showy males have stronger, healthier, more successful offspring.

OTHER FACTORS CAN CHANGE ALLELE FREQUENCIES ACROSS GENERATIONS

Natural selection is not the only process that can change allele frequencies, and so change traits across generations. Random factors also can change populations. *Genetic drift* is the general term for random changes in allele frequencies. To understand genetic drift, you must return to the idea of probabilities. Any population contains many **alleles** for each gene and many possible allele combinations. When the individuals of one generation mate and produce offspring, there is some probability that a particular allele will, or will not, appear in the next generation, just by chance. Because of the random chances involved in the formation of gametes and in fertilization, not every egg and sperm will end up contributing to a zygote and developing into a new individual. If an allele is lost in a population for random reasons, genetic drift has occurred. Or if an allele becomes more frequent in a population for random reasons, genetic drift has occurred. Genetic drift does not act to match a population to its environment. The changes in allele frequencies that occur as a result of genetic drift may be helpful to the survival of the population, or they may not be helpful to the survival of a population, but they are not the result of selection.

PROCESSES OF SPECIATION[7]

Speciation is the production of new species. It demands mechanisms for bringing new species into existence and mechanisms for maintaining them and building them into cohesive units of interbreeding individuals that maintain some degree of isolation and individuality. There is a threshold at which *microevolution* (evolution through

---

7   Richard J. Huggett, "Biogeographical Processes I: Speciation, Diversification, and Extinction," *Fundamentals of Biogeography*, pp. 18–24. Copyright © 2004 by Taylor & Francis Group. Reprinted with permission.

adaptation within species) becomes *macroevolution* (evolution of species and higher taxa). Once this threshold is traversed, evolutionary processes act to uphold the species' integrity and fine-tune the new species to its niche: gene flow may smother variation; unusual genotypes may be less fertile, or may be eliminated by the environment, or may be looked over by would-be mates. Various mechanisms may thrust a population through the *speciation threshold*, each being associated with a different model of speciation: allopatric speciation, peripatric speciation, stasipatric speciation, and sympatric speciation (Figure 5.28). Evolutionary biologists argue over the effectiveness of each type of speciation (e.g. Losos and Glor 2003).

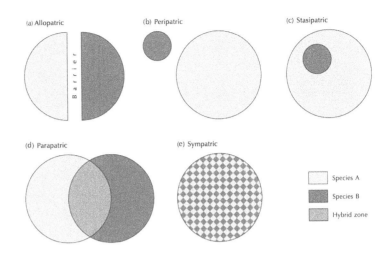

Figure 5.28: Types of speciation: allopatric speciation, peripatric speciation, stasipatric speciation, parapatric, and sympatric speciation.

ALLOPATRIC SPECIATION

Geographical isolation reduces or stops gene flow, severing genetic connections between once interbreeding members of a continuous population. If isolated for long enough, the two daughter populations will probably evolve into different species. This mechanism is the basis of the classic model of **allopatric** ('other place' or geographically separate) **speciation**, as propounded by Ernst Mayr (1942) who called it *geographical speciation* and saw geographical subdivision as its driving force. Mayr recognized three kinds of allopatric speciation: strict allopatry without a population bottleneck; strict allopatry with a population bottleneck; and extinction of intermediate populations in a chain of races:

1. *Strict allopatry without a population bottleneck* occurs in three stages. First, the original population extends its range into new and unoccupied territory. Second, a geographical barrier, such as a mountain range, forms and splits the population into two; and genetic modification affects the separate populations to the extent that if they come back into contact, genetic isolating mechanisms will prevent their reproducing. The model includes cases where species extend their range by traversing an existing barrier, as when birds cross the sea to colonize an island.

2. *In strict allopatry with a population bottleneck,* a small band of founder individuals (or even a single gravid female) colonizes a new area. Mayr argued that the founding population would carry but a small sample of the alleles present in the parent population, and the colony would have to squeeze though a genetic bottleneck; he called this the *founder effect*. However, from the 1970s onwards, it became apparent that the bottleneck might not be as tight as originally supposed, even with just a few founding intervals (White 1978, 109). Some 10 to 100 founding colonists may carry a substantial portion of the alleles present in the parent population. Even a single gravid female colonist, providing she is heterozygous at 10–15 per cent of her gene loci and providing an equally heterozygous male (largely of different alleles) fertilized her, could carry a considerable amount of **genetic variability**. Admittedly, a newly founded colony will initially be more homozygous and less polymorphic than the parent population. If the colony should survive, gene mutation should restore the level of polymorphism, probably with new alleles with new allele frequencies.

3. Two species of European gull—the herring gull (*Larus argentatus*) and the lesser black-backed gull (*L. fuscus*)—exemplify the extinction of intermediate populations in a chain of species (Figure 5.29). These species are the terminal members of a chain of *Larus* subspecies encircling the north temperate region. Members of the chain change gradually but the end members occur sympatrically in northwest Europe without **hybridization**.

Vicariance events and dispersal-cum-founder events may drive allopatric speciation (Figure 5.30). Two species of North American pines illustrate vicariance speciation. Western North American lodgepole pine (*Pinus contorta*) and eastern North American jack pine (*P. banksiana*) evolved from a common ancestral population that the advancing Laurentide Ice Sheet split asunder some 500,000 years ago. The colonization of the Galápagos **archipelago** from South America by an ancestor of the present giant tortoises (*Geochelone* spp.), probably something like its nearest living relative the Chaco tortoise (*G. chilensis*), is an example of a dispersal and founder event. The giant tortoises are all the same species on the Galapagos—*G. elephantopus*—but there are 11 living subspecies and 4 extinct subspecies. Five subspecies occur on Isabela and the rest occur on different islands.

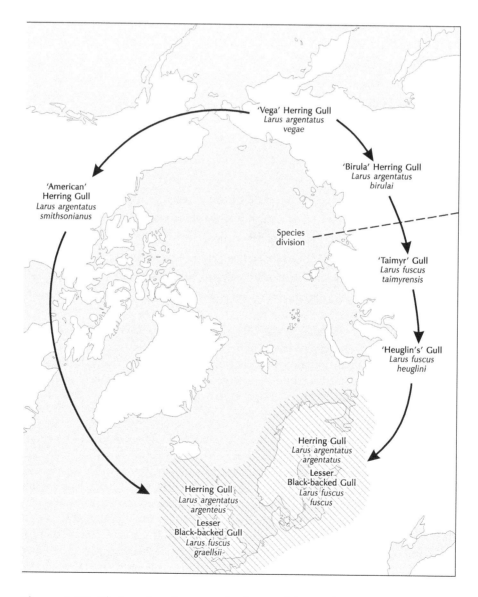

Figure 5.29: Chain or ring of species of gull around the North Pole. Change between subspecies is gradual but the two end members live sympatrically in northwest Europe.

Once populations are isolated and become differentiated, they may then stay isolated and never come into contact again. In this case, it may take a long time for reproductive isolation to occur and it is difficult to know when the two populations are different species. If contact is reestablished, perhaps because the barrier disappears, then three things may happen: (1) the populations may not interbreed, or fail to produce fertile offspring, reproductive isolation is complete, and speciation has occurred. This happened with the kaka and kea in New Zealand. (2) The two populations may hybridize, but the hybrids may be less fit than the offspring of the within-populations matings. *Reinforcement* is the process that selects within-population matings, and *isolating mechanisms* are traits that evolve to augment reproductive isolation. (3) The two populations may interbreed comprehensively, producing fertile fit hybrids, so that the populations merge and the differentiation is diluted, and eventually disappears.

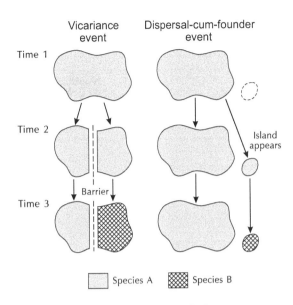

Figure 5.30: The chief drivers of allopatric speciation: vicariance events and dispersal-cum-founder events. *Source: Adapted from Brown and Lomolino (1998)*

PERIPATRIC SPECIATION

This is a subset of allopatric speciation. **Peripatric speciation** occurs in populations on the edge (perimeter) of a species range that become isolated and evolve divergently to create new species. A small founding population is often involved. An excellent example of this is the paradise kingfishers (*Tanysiptera*) of New Guinea (Mayr 1942). The main species, the common paradise kingfisher (*T. galatea galatea*), lives on the main island. The surrounding coastal areas and islands house a legion of morphologically distinct races of the paradise kingfishers (Figure 5.31).

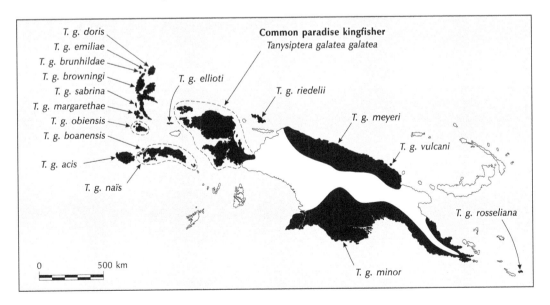

Figure 5.31: Races of paradise kingfishers (*Tanysiptera*) on New Guinea and surrounding islands. *Source: Partly adapted from Mayr (1942)*

Table 5.2 Living Lemurs

GENUS	NUMBER OF SPECIES IN GENUS	SPECIES
Woolly lemurs (*Allocebus*)	2	Hairy-eared dwarf lemur (*Allocebus trichotis*)
		Eastern woolly lemur (*Avahi laniger*)
Western woolly lemur (*Avahi*)	1	Western woolly lemur (*Avahi occidental's*)
Dwarf lemurs (*Cheirogaleus*)	2	Greater dwarf lemur (*Cheirogaleus major*)
		Fat-tailed dwarf lemur (*Cheirogaleus medius*)
Aye-ayes	1	Aye-aye (*Daubentonea madagascariensis*)
		[One extinct species]
True lemurs (*Eulemur*)	5	Crowned lemur (*Eulemur coronatus*)
		Brown lemur (*Eulemur fulvus*)
		Black lemur (*Eulemur macaco*)
		Mongoose lemur (*Eulemur mongoz*)
		Red-bellied lemur (*Eulemur rubriventer*)
Gentle lemurs (*Hapalemur*)	3	Golden bamboo lemur (*Hapalemur aureus*)
		Grey gentle lemur (*Hapalemur griseus*)
		Broad-nosed gentle lemur (*Hapalemur simus*)
Indris (*Indri*)	1	Indri (*Indri indri*)
Ring-tailed lemurs (*Lemur*)	1	Ring-tailed lemur (*Lemur catta*)
Sportive lemurs (*Lepilemur*)	7	Grey-backed sportive lemur (*Lepilemur dorsalis*)
		Milne-Edwards' sportive lemur (*Lepilemur edwardsi*)
		White-footed sportive lemur (*Lepilemur leucopus*)
		Small-toothed sportive lemur (*Lepilemur microdon*)
		Weasel sportive lemur (*Lepilemur mustelinus*)
		Red-tailed sportive lemur (*Lepilemur ruficaudatus*)
		Northern sportive lemur (*Lepilemur septentrionalis*)
Mouse lemurs (*Microcebus*)	4	Grey mouse lemur (*Microcebus murinus*)
		Pygmy mouse lemur (*Microcebus myoxinus*)
		Golden-brown mouse lemur (*Microcebus ravelobensis*)
		Red mouse lemur (Microcebus rufus)
Mirza	1	Coquerel's dwarf lemur (*Mirza coquereli*)
Phaner	1	Fork-crowned dwarf lemur (*Phaner furcifer*)
Sifakas (*Propithecus*)	3	Diademed sifaka (*Propithecus diadema*)
		Tattersall's sifaka (*Propithecus tattersalli*)
		Verreaux's sifaka (*Propithecus verreauxi*)
Ruffed lemurs (*Varecia*)	1	Ruffed lemur (*Varecia variegata*)
		Ring-tailed lemurs (Lemur)
		Sportive lemurs (*Lepilemur*)
		[One recently extinct species]

PARAPATRIC SPECIATION

Parapatric (abutting) speciation is the outcome of **divergent evolution** in two populations living geographically next to each other. The divergence occurs because local adaptations create genetic gradients or clines. Once

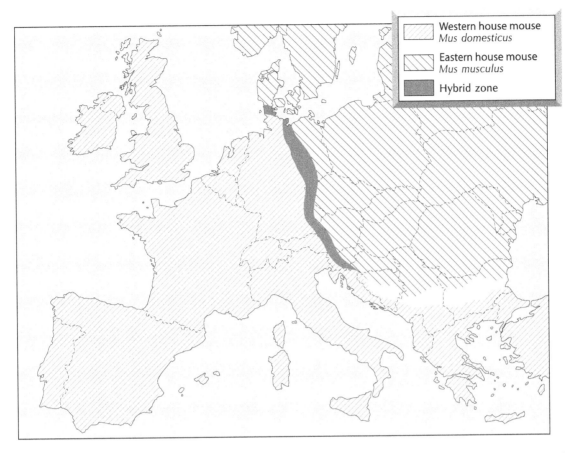

Figure 5.32: The distribution of the light-bellied eastern house mouse (*M. musculus*) of eastern Europe and the dark-bellied western house mouse (*M. domesticus*) of western Europe (which some authorities treat as a subspecies of *M. musculus*) and the zone of hybridization between them. *Source: Adapted from distribution maps in Mitchell-Jones et al. (1999)*

established, a cline may reduce gene flow, especially if the species is a poor disperser, and selection tends to weed out hybrids and increasingly pure types that wander and find themselves at the wrong end of the cline. A true hybrid zone may develop, which in some cases, once reproductive isolation is effective, will disappear to leave two adjacent species. An example is the main species of the house mouse in Europe (Hunt and Selander 1973). A zone of hybridization separates the light-bellied eastern house mouse (*Mus musculus*) of eastern Europe and the dark-bellied western house mouse (*M. domesticus*) of western Europe (Figure 5.32).

SYMPATRIC SPECIATION

Sympatric speciation occurs within a single geographical area and the new species overlap—there is no spatial separation of the parent population. Separate genotypes evolve and persist while in contact with each other. Once deemed rather uncommon, new studies suggest that parapatric and sympatric speciation may be a potent process of evolution (e.g. Via 2001).

Several processes appear to contribute to sympatric speciation. Disruptive selection favours extreme phenotypes and eliminates intermediate ones. Once established, natural selection encourages reproductive isolation through *habitat selection or positive assertive mating* (different phenotypes choose to mate with their own kind). Habitat selection in insects may have favoured sympatric speciation and account for much of the large diversity of that group. *Competitive selection*, a variant of disruptive selection, favours phenotypes within a species that avoid intense competition and clears out intermediate types.

Two irises that grow in the southern USA appear to result from sympatric speciation. The giant blue iris (*Iris giganticaerulea*) grows in damp meadows, while its close relative, the copper iris (*I. fulva*), grows in drier riverbanks. The two species hybridize, but the hybrids are not as successful in growing on either the dry or very wet sites as the pure forms and usually perish. So, the inability of hybrids to survive restricts the gene flow between the two species. ...

CONVERGENT EVOLUTION AND PARALLEL EVOLUTION[8]

Convergent evolution is the process whereby different species independently evolve similar traits as a result of similar environments or selection pressures. For instance, sharks and bony fish, whales and dolphins, and extinct sea-going reptiles (ichthyosaurs) all evolved streamlined, torpedo-shaped bodies for cutting through water.

Parallel evolution (or **parallelism**) refers to changes in two closely related stocks that differ in minor ways and that both go through a similar series of evolutionary changes. It is similar to convergence, except that in convergence the original species are from very different stocks, unlike the stocks in parallel evolution, which are similar to start with. Marsupials and placental mammals are a case in point, though sometimes thought of as a case of convergent evolution.

SPECIES DEATH: EXTINCTION[9]

Extinction is the doom of the vast majority of species (or genera, families, and orders); it is the rule, rather than the exception. A local extinction or extirpation is the loss of a species or other taxon from a particular place, but other parts of the gene pool survive elsewhere. The American bison (*Bison bison*) is now extinct over much of its former range, but survives in a few areas. A **global extinction** is the total loss of a particular gene pool. When the last dodo died, its gene pool was lost forever. Supraspecific groups may suffer extinctions. An example is the global extinction of the sabre-toothed cats, one of the main branches of the cat family. A **mass extinction** is a catastrophic loss of a substantial portion of the world's species. Mass extinctions stand out in the fossil record as times when the extinction rate runs far higher than the background or normal extinction rate. Some 99.99 per cent of all extinctions are normal extinctions.

The 'life-expectancy' of species varies between different groups. The fossil record suggests that mammal genera last about 10 million years, with primate genera enduring only 5 million years. Individual species survive even less

---

8   Richard J. Huggett, "Biogeographical Processes I: Speciation, Diversification, and Extinction," *Fundamentals of Biogeography*, pp. 30-31. Copyright © 2004 by Taylor & Francis Group. Reprinted with permission.

9   Richard J. Huggett, "Biogeographical Processes I: Speciation, Diversification, and Extinction," *Fundamentals of Biogeography*, pp. 32-34. Copyright © 2004 by Taylor & Francis Group. Reprinted with permission.

time, something around 1 to 2 million years for complex animals. On the other hands, 'living fossils' appear to have persisted for ages with little change. Examples are the horseshoe crab (*Limulus* spp.), a relative of the spiders, which has lived and changed little for at least 300 million years; cycads, which are 'living fossil' plants surviving from the **Mesozoic era**; and the ginkgo, which is remarkably similar to specimens that lived around 100 million years ago (Zhou and Zheng 2003). Probably the most famous 'living fossils' are the coelacanths—*Latimeria chalumnae* was found in 1938, *L. menadoensis* in 1998, and an as yet unnamed species in 2000. Coelacanths have persisted nearly unchanged for 70 million years.

Periods of rapid climatic change, sustained volcanic activity, and asteroid and comet impacts seem to cause mass extinctions. Normal extinctions depend on many interrelated factors that fall into three groups—biotic, evolutionary, and abiotic.

BIOTIC FACTORS

Most biotic factors of extinction are **density-dependent factors**. This means their action depends upon population size (or density). The larger the population, the more effective is the factor. Density-dependent factors are chiefly biotic in origin. They include factors related to biotic properties of individuals and populations (body size, niche size, range size, population size, generation time, and dispersal ability) and factors related to interactions with other species (competition, disease, parasitism, predation).

BIOTIC PROPERTIES

Body size, niche size, and range size all affect the probability of extinction. As a rule, large animals are more likely to become extinct than small animals. Smaller animals can probably better adapt to small-scale habitats when the environment changes. Large animals cannot so easily find suitable habitat or food resources and so find it more difficult to survive. Specialist species with narrow niches are more vulnerable to extinction than are generalists with wide niches.

Small populations are more prone to extinction through chance events, such as droughts, than are large populations. In other words, there is safety in numbers. Tropical birds living in patches of Amazon forest show that populations of 50 or more are about 5 times less likely to go extinct locally than are populations of 5 or fewer. Species with rapid generation times stand more chance of dodging extinction. Good dispersers are better placed to escape extinction than poor dispersers, as are species with better opportunities for dispersal. In addition, a species with a large gene pool may be better able to adapt to environmental changes than species with a small gene pool.

Geography can be important—widespread species are less likely to go extinct than species with restricted ranges. This is because restricted range species are more vulnerable to chance events, such as a severe winter or drought. In a widespread species, severe events may cause local extinctions but are not likely to cause a global extinction. This generalization is borne out by defaunation experiments on red mangrove (*Rhizophora mangle*) islands in the Florida Keys, USA, where the insects, spiders, mites, and other terrestrial animals were exterminated with methyl bromide gas (see Simberloff and Wilson 1970). Analysis of the data revealed that the probability of invertebrate extinctions decreased with the number of islands occupied (Hanski 1982). It should be pointed out that a widespread distribution is not a guarantee of extinction avoidance. The passenger pigeon (*Ectopistes migratorius*) and the American chestnut (*Castanea dentata*) were abundant with widespread distributions in eastern North America in the nineteenth century, but suffered *range collapse* and extinction in the case of the passenger pigeon and near extinction in the case of the American chestnut within 100 years. The bison (*B. bison*), trumpeter swan (*Olor buccinator*), whooping crane (*Grus americana*), and sandhill crane (*G. canadensis*), also once widely distributed in North America, have suffered range collapses.

Widespread species also appear to be less at risk than restricted species to mass extinctions. For instance, extinction rates of marine bivalves and gastropods that lived along the Atlantic and Gulf coastal plains of North America in the late Cretaceous period increased with species range (Jablonski 1986).

BIOTIC INTERACTIONS

Competition can be a potent force of extinction. Species have to evolve to outwit their competitors, and a species that cannot evolve swiftly enough is in peril of becoming extinct.

Virulent pathogens, such as viruses, may evolve or arrive from elsewhere to destroy species. The fungus *Phiostoma ulmi*, which is carried mainly by the Dutch elm beetle (*Scolytus multistriatus*), causes Dutch elm disease. Starting in the Netherlands, Dutch elm disease spread across continental Europe and into the USA during the 1920s to 1940s, ravaging the elm populations. After a decline in Europe (but not in the USA), it re-emerged as an even more virulent form (described as a new species—*Ophiostoma novoulmi*) in the mid-1960s to affect Britain and most of Europe.

Predators at the top of food chains are more susceptible to a loss of resources than are herbivores lower down. A chief factor in the decline of tigers is not habitat loss or poaching, but a depletion of the ungulate prey base throughout much of the tigers' range (Karanth and Stith 1999).

Island mammal, bird, and reptile populations are especially vulnerable to all sorts of competitive and predatory introduced species (Table 5.3). Since 1600 (and up to the late 1980s), 113 species of birds have become extinct. Of this total, 21 were on mainland areas and 92 on islands (Reid and Miller 1989). In many cases, numerous species of sea birds survive only on outlying islets where introduced species have failed to reach. The story for mammals and reptiles is similar.

EVOLUTIONARY FACTORS

Several evolutionary changes may, by chance, lead to some species being more prone to extinction than others. *Evolutionary blind alleys* arise when a loss of genetic diversity during evolution fixes species into modes of evolutionary development that become lethal. A species may evolve on an island and not possess the dispersal mechanisms to escape if the island should be destroyed or should experience climatic change. Some species may become overspecialized through adaptation and fall into *evolutionary traps*. Faced with environment change, overspecialized species may be unable to adapt to the new conditions, their over-specialization serving as a sort of evolutionary straitjacket

Table 5.3 Recorded Extinction of Mammals, Birds, and Reptiles, 1600 to 1983

TAXON	MAINLAND[a]	ISLAND[b]	OCEAN	TOTAL EXTINCTIONS	APPROXIMATE NUMBER OF SPECIES IN TAXON	PERCENTAGE OF TAXON LOST
Mammals	30	51	4	85	4,000	2.1
Birds	21	92	0	113	9,000	1.3
Reptiles	1	20	0	21	6,300	0.3

Source: Adapted from Reid and Miller (1989)
Notes: a Landmasses 1 million km² (the size of Greenland) or larger. b Landmasses less than 1 million km²

that keeps them 'trapped.' An interesting upshot of this idea is that species alive today must be descendants of non-specialized species. Behavioural, physiological, and morphological complexity, as varieties of specialization, also appear to render a species more prone to extinction. Simple species—marine bivalves for example—survive for about 10 million years, whereas complex mammals survive for 3 million years or less.

ABIOTIC FACTORS

Abiotic factors of extinction are usually density-independent factors, which means that they act uniformly on populations of any size. Density-independent factors tend to be physical in origin—climatic change, sea-level change, flooding, asteroid and comet impacts, and other catastrophic events. These factors often produce fluctuations in population size that can end in extinction. Take the example of the song thrush (*Turdus philomelos*). This bird lives throughout the British Isles except Shetland (Venables and Venables 1955). It was absent from Shetland in the nineteenth century but established a colony on the island in 1906, breeding near trees, which were scarce in Shetland, the largest group being planted in 1909. By the 1940s, about 24 of breeding pairs inhabited the island. The severe winter of 1946–7 reduced the population to some three or four pairs from then until 1953. Somewhere between 1953 and 1969 the Shetland's song thrushes died out.

Abiotic factors are usually implicated in mass extinctions. However, several researchers stress the potential role of diseases as drivers of mass extinctions. Lethal pathogens carried by the dogs, rats, and other animals associated with migrating humans may have caused the Pleistocene epoch mass extinctions (MacPhee and Marx 1997). Similarly, it is possible that the terminal Cretaceous extinction event might have resulted from changes of palaeogeography, in which land connections created by falling sea-levels allowed massive migrations from one landmass to another, leading to biotic stress in the form of predation and disease:

> The shallow oceans drained off and a series of extinctions ran through the saltwater world. A monumental immigration of Asian dinosaurs streamed into North America, while an equally grand migration of North American fauna moved into Asia. In every region touched by this global intermixture, disasters large and small would occur. A foreign predator might suddenly thrive unchecked, slaughtering virtually defenseless prey as its population multiplied beyond anything possible in its home habitat. But then the predator might suddenly disappear, victim of a disease for which it had no immunity. As species intermixed from all corners of the globe, the result could only have been global biogeographical chaos.

(Bakker 1986, 443)

EDITOR'S CONCLUSION TO BIOGEOGRAPHICAL PROCESSES BY RICHARD HUGGETT

A by-product of recent advances in the study of mass extinctions is the observation that they are typically followed by periods of adaptive radiation. In other words, mass extinctions not only wipe out old species, but also vacate ecological niches, allowing new species to evolve to fill them. These new species tend to be the descendants of former "underdogs," those that adapted to the peripheries of their ecosystems prior to mass extinctions. Such an existence would likely fragment these peripheral species into smaller, isolated populations that undergo allopatric speciation. Meanwhile, the dominant species, those that became well established and stable in their respective niches, would be the most homogenous in form and habit and dependent on environmental stasis. A sudden change in the environment, say one that results in a mass extinction, would most negatively affect the main players in the ecosystem, whereas the outliers, those "toughened up" by having to hide in the shadows of other species, would be better equipped for change. A prime example of this is the diversification of mammals in the Cenozoic following the extinction of the non-avian dinosaurs.

Figure 6.1: A view of the Earth from above the moon's surface. From the moon's perspective, the Earth doesn't rise or set.

LIFE BEGINS

Scientists don't know exactly what led to the formation of our particular solar system. Abundant telescopic observations of star and planet formation in other areas of our galaxy tell us that our solar system likely formed from the **accretion** of nebulous gases and dust. During accretion, materials in space clump together because of gravity. This clumping happens rather slowly over millions of years, bit by bit, until these balls of gas and dust reach the sizes of stars and planets.

Theories of the origin of the universe suggest that the earliest stars to form accreted from nebulae of hydrogen and helium, which are very small, simple elements. Only in the cores of stars can hydrogen and helium undergo nuclear fusion and create heavier elements. The abundance of heavy elements in our own solar system implies the cloud of material preceding it was already rich in heavy elements. This could only be the case if this cloud of material had formed from the supernova explosion of a previous star. If that is true, then nearly every bit of matter in our solar system, including us, was generated in the core of a defunct star.

Our solar system, therefore, likely began as a cloud of dust and gas, a mixture of all the bits of matter that now compose the sun and planets. Several stages of development ensued once this cloud began to collapse in on itself due to gravity.

1. The cloud shrinks: Gravity causes the matter in the cloud to **condense**, or pull together more tightly. This increases the pressure and temperature in the cloud, especially in the middle. As the cloud condenses, it spins. The spin may be due to the motion of the original cloud, or possibly it originates from complex gravitational interactions as the cloud collapses. This assumed spinning best explains the turning of our sun on its own axis as well as the orbiting pattern of the planets.

2. Our sun forms: The material in the center of the cloud forms a sphere of superheated gas that reaches the pressures and temperatures needed for nuclear fusion to initiate in its core. This fusion produces huge amounts of energy in the process (solar radiation) that radiates outward.

3. A rotating disc of material surrounds the sun: Once the sun is born, the other planets still need to form. **Solar wind**, a "breeze" of energy and charged particles from the sun, blows outward from the middle of this **circumstellar disc** (Figures 6.2 and 6.3) and causes **differentiation**. During differentiation, the solar wind blows the more lightweight dust and gases toward the outskirts of the disc, and the heavier elements become more concentrated nearer the sun as a result.

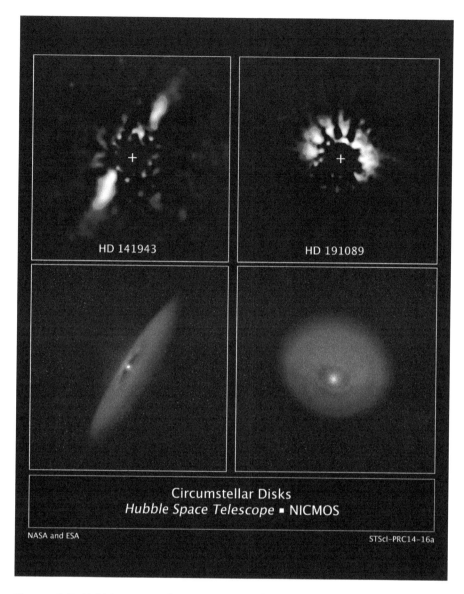

Figure 6.2: Hubble images of two stars surrounded by circumstellar discs. The bottom photos are illustrations to better show the orientations of the disks relative to Earth.

4. The planets form: After the solar wind differentiates the circumstellar disc, the planets accrete at various distances from the sun. The circumstellar disc begins to thin out as these initially small clumps grow in size and collect dust and gas along their orbital paths circling the new sun. As time passes, the "dust clears" as the planets become massive enough to effectively clear their paths around the sun of dust and debris. The final stages of this process result in a **period of heavy bombardment.** This refers to the period of time (lasting until about 3.8 billion years ago) during which the planets had already formed but were still being bombarded by numerous asteroids that composed the last little bits of the circumstellar disc. Planets and moons with no atmosphere have virtually no erosion and therefore still display numerous craters on their surfaces left over from the period of heavy bombardment (Figure 6.4).

5. The planets differentiate: Once formed, the planets were still very hot from pressures and temperatures exerted on them during accretion as well as the heavy bombardment of asteroids. This kept Earth, for example, in a molten state, and during this time, the heavy metals sank to the bottom, the gases rose to the surface, and the Earth, as a whole, differentiated so that now it has a core, mantle, and crust. This occurred with all the planets roughly at the same time and a little after the sun had begun to radiate energy.

Figure 6.3: An illustration of a circumstellar disc in the process of accretion.

Figure 6.4: Ancient craters on the moon left behind by asteroid impacts during the early stages of the solar system. The main crater in the middle is called Daedalus.

By 3.8 billion years ago, the Earth had cooled enough to allow a somewhat stable crust to form, and the first forms of life came about slightly thereafter, as indicated by the fossil record.

THE ARCHEAN EARTH

All of these important events took place during the **Archean eon** (4.6–2.5 billion years ago). Few rocks remain from this period of time, which makes this history speculative. The earliest minerals on Earth have ages about 4.4–4.0 billion years ago and are found in old fragments of crust, such as northern Canada, Africa and Australia. These tiny crystals indicate the Earth had an early, rudimentary crust during this time.

The Earth was hotter then than it is today. Much of the original heat left over from accretion was still radiating from the Earth, and the radiogenic heat from the core was stronger, making Earth highly volcanic during this time. The early lithosphere would have had much greater rates of tectonic movement than today, and there would have been a lot of earthquakes and volcanoes as a result.

As the Earth cooled, the lithosphere continued to evolve so that by the end of the Archean, at least of a third of our present volume of continental crust had formed along the margins of subduction zones.

HOW THE ATMOSPHERE FORMED

The cloud of dust and gas the Earth formed from would have been rich in hydrogen and helium, and the earliest atmosphere the Earth would have collected through accretion would have also been made mostly of these lightweight gases. The small mass of hydrogen and helium makes is easier for them to be blown into space for two reasons: One, they

Figure 6.5: False-color cathodoluminescence image of a 400-micrometer zircon. Conventional U-Pb geochronology shows this is the oldest known (4.4 Ga, billion years) concordant zircon from Earth. (credit: John Valley, http://www.news.wisc.edu/newsphotos/zircon1.html. Copyright © by John Valley. Reprinted with permission.)

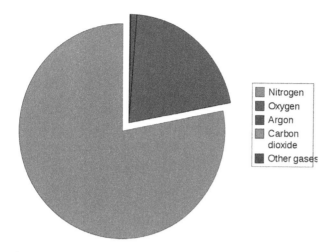

Figure 6.6: The composition of our modern atmosphere is dominated by nitrogen and oxygen.

would only have a weak gravitational attraction to the Earth, and two, the Earth's magnetic field had not fully developed, and as such, the atmosphere would have been more vulnerable to being blown away by solar winds.

Our atmosphere today is a mixture of gases, mostly nitrogen and oxygen, with all the other gases making up about 2 percent of air (Figure 6.6). Once the Earth's magnetic field stabilized, the atmosphere accumulated through the outgassing of volcanoes. Volcanic gases include but are not limited to water vapor, carbon monoxide, carbon dioxide, sulfur dioxide, nitrogen, chlorine, and hydrogen. These gases were belched out of volcanoes and remained as part of the atmosphere due to gravity.

One gas that is *not* emitted in large amounts from volcanoes is free oxygen (O_2), which would have been very uncommon in the early atmosphere. Instead, oxygen was mostly bound to other elements in volcanic gases (e.g., CO_2, H_2O). To a small degree, the splitting of molecules such as these by solar radiation would have released minute amounts of free oxygen, but the Earth didn't see large amounts of it produced until about 2.3 billion years ago, after organisms had begun to photosynthesize and effectively strip free oxygen from carbon dioxide.

HOW THE OCEANS FORMED

Mineralogic evidence suggests oceans began to accumulate in the Early Archean, perhaps not long after the formation of the crust. Scientists do not know what the primary source of our water was but agree it was a combination of water vapor from volcanoes and water from icy asteroids and comets that accreted to Earth during and after its formation. The Earth is not the only body in the solar system with water; most planets and moons have it. The average temperature of our planet, however, keeps the water from entirely evaporating away or from soaking into the ground and freezing solid. It is mostly liquid and sits atop the crust, making it very accessible to life on the planet. The ongoing outgassing of water vapor from volcanoes continues to add small amounts of water to the oceans and reflects continuous, ongoing differentiation of our planet.

ABIOGENESIS AND LIVING THINGS

Abiogenesis is a theory that suggests the first living things on Earth were created through spontaneous chemical reactions at the Earth's surface. A living thing is informally defined as something that grows through metabolization and is capable of reproduction.

A true scientific definition of life, however, does not exist; in fact, life is not technically a scientific word at all. NASA, for instance, participates in research that seeks out life on other moons and planets of the solar system, yet they have no concrete definition of life, only the notion that it is more or less a "self-sustaining system capable of Darwinian evolution" (NASA).

It is important to point out that the theory of evolution does not attempt to address the origins of life; it only serves as a proposed mechanism by which species arise and change through time. Abiogenesis and evolution should not be confused with each other, nor should they be considered dependent on one another to exist.

Early scientific thoughts about the origins of life focused on the idea of **spontaneous generation** (Figure 6.7). Great thinkers, such as Aristotle, subscribed to the once-popular notion that piles of hay sprout mice and that insects arise from putrid matter. It was thought that maggots form from rotting meat and beetles form directly from cow dung. By the mid-1800s, this theory had been mostly disproved. By the mid-1900s, scientists began to hypothesize that living things arose from chemical reactions in the early Earth that created progressively more complex organic compounds. This theory was tested in 1953 in the famous Miller-Urey experiment (Figure 6.8).

Figure 6.7: Early ideas such as spontaneous generation suggested that fruit flies, for example, were created somehow from the fruit itself.

THE MILLER-UREY EXPERIMENT

Inspired by a hypothesis by Alexander Oparin of Russia and J.B.S. Haldane of Britain (which stated that the conditions of the primitive Earth were favorable for building complex organic molecules), Stanley Miller and

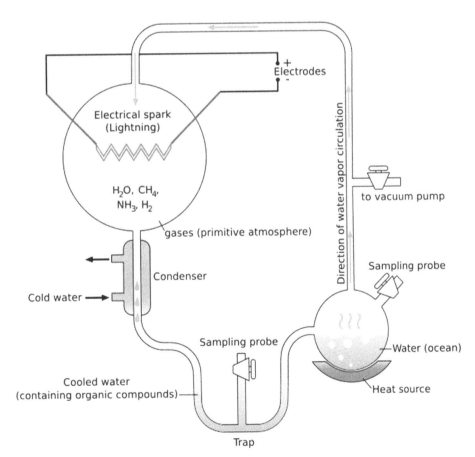

Figure 6.8: The Miller-Urey Experiment found that simple organic compounds could be created through abiologic reactions of inorganic substances.

Harold Urey of the University of Chicago designed an experiment that recreated the Earth's early atmosphere to see if these compounds could be inorganically produced. The theory is that as long as the appropriate chemical ingredients are around and some form of energy can be applied to them, reactions could be catalyzed that form organic compounds.

The experiment consisted of a network of glass pipes and flasks that pumped water through a vessel containing a mixture of atmospheric gases (methane, ammonia, water vapor, and hydrogen). The gases were sparked with electricity to simulate lightning strikes in the Earth's early atmosphere. After a few days, the water became cloudy with a primitive mixture of all twenty amino acids found in living things today. Although this experiment was successful in that it showed that somewhat simple organic compounds (monomers) can be produced from gases, it did not produce the more complex nucleic acids such as RNA and DNA, which are polymers. Furthermore, it has been determined more recently that the mixture of gases used in the experiment was an incorrect representation of the early atmosphere. Subsequent results from this experiment that utilize a more correct mixture of gases have not been as promising as the originals.

THE "PRIMORDIAL SOUP"

This term was phrased by J.B.S. Haldane to describe the accumulations of simple monomers such as amino acids in environments such as beaches, mud, and shallow ponds. A large hurdle in the understanding of how life began is the way that these monomers could polymerize into larger, more complex organic molecules. An issue with water is that it tends to prohibit the linking up of monomers into more complex polymers. However, when monomers such

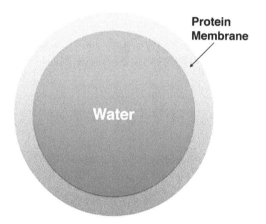

Figure 6.9: Protobionts (thermal proteins) are among the most complex organic compounds created in the lab. They form self-replicating spheres, but are not actually alive.

as these are concentrated and subsequently heated up and dried out, they have been observed to link up and form *thermal proteins*, which are chains of up to two hundred amino acids. In terms of complexity, thermal proteins, also called **protobionts** (Figure 6.9), are about halfway between inorganic compounds and living tissue. These have been observed to spontaneously reorganize themselves into membrane-bound microspheres that periodically divide just like cells do. Perhaps the formation of thermal proteins occurred along the shores of volcanic islands, along deep-sea hydrothermal vents (Figure 6.10), or within dried-up puddles, where monomers became concentrated, heated and dried by volcanic heat and/or solar radiation.

EXTRATERRESTRIAL ORIGINS

In 1969, residents of Victoria, Australia, witnessed a fireball streak across the sky and break apart into over one hundred kilograms of individual fragments that rained down over an area of about thirteen square kilometers. Analysis of the meteoritic fragments, collectively named the Murchison meteorite based on the town nearby, showed that the rocks were primitive in composition (representing material from the early solar system) and had undergone some form of alteration from water present on the host asteroid from which they came. As these fragments were collected within days of landing, the possibility of terrestrial contamination was small, and more

Figure 6.10: A) Hydrothermal vent at the Eifuku volcano. Locations like these may have been where the first life was formed; B) Other hypotheses suggest that the building blocks of life came to earth via asteroids. In this photo, a large meteor exploded over Chelyabinsk City, Russia. Over 1,000 people were injured from the shock wave.

recent studies have further concluded that molecules from the meteorite are indeed of extraterrestrial origin (Figure 6.11).

Within the meteorites, scientists discovered a mixture of at least fifteen amino acids and a plethora of other compounds, many of which had been synthesized in the lab in the Miller-Urey experiment. In fact, the complexity of organic compounds in the meteorite rivals that of petroleum. This discovery prompted scientists to explore the possibility that the polymerization of monomers could have taken place within the early solar system, possibly even before the planets had fully formed. This is a useful hypothesis, as studies continue to show that the early conditions on Earth may not have been sufficient to create any compounds more complex than protobionts.

This theory suggests that while the circumstellar disc was still thinning out and accreting into planets, intense UV light from the sun and other stars would have irradiated nebular material such as gas-rich chunks of ice with energy, catalyzing the reactions. Experiments and models show that these conditions would have been favorable to the production of organic molecules, which, since 1969, have been discovered in other meteorites, asteroids, and comets.

It is possible, then, that the polymerization of amino acids and other compounds occurred in space, and these organic molecules rained down onto the newly formed planets, coating their surfaces with the building blocks of DNA and RNA. As of today, experiments have failed to show how the final steps of DNA and RNA production were achieved.

Figure 6.11: A powdered sample of the Murchison meteorite extracted from the fragment for analysis.

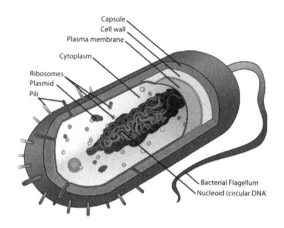

Capsule
Cell wall
Plasma membrane
Cytoplasm
Ribosomes
Plasmid
Pili
Bacterial Flagellum
Nucleoid (circular DNA)

Figure 6.12: A prokaryotic cell, like those that evolved close to four billion years ago.

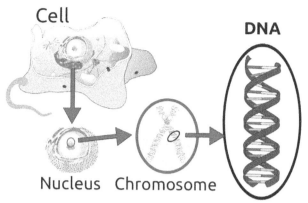

Cell

DNA

Nucleus Chromosome

Figure 6.13: A eukaryotic cell is more complex than a prokaryote.

THE OLDEST KNOWN ORGANISMS

The earliest organisms detected in rocks are bacteria and archaea. Archaea are like bacteria but differ in their genetic makeup and are capable of surviving in more extreme environments than bacteria. Both bacteria and archaea are considered to be **prokaryotes** (Figure 6.12), which are single-celled organisms that lack a membrane around their nuclei as well as other organelles present in the more complex **eukaryotic cells** (Figure 6.13) that came later.

Chemical evidence of the first prokaryotes comes from rocks about 3.8 billion years old, and by 3.5 billion years ago, photosynthesizing (Figure 6.14) bacteria, called **cyanobacteria** (Figure 6.15), were beginning to form layered colonies called **stromatolites** (Figure 6.16) along ancient shorelines. Prior to photosynthesis acting as a metabolic process, the earliest prokaryotes were likely heterotrophs, organisms that cannot produce their own nutrients. Instead, these early single-celled organisms probably extracted phosphate from water or rocks and combined it with various gases to produce ATP (adenosine triphosphate), which provides the energy used to drive chemical reactions in cells.

Shortly thereafter, prokaryotes likely began to use fermentation as an autotrophic mechanism to produce nutrients, specifically by the splitting of sugars to release energy. Fermentation does not require an oxygen-rich environment (which would not have been present on Earth at the time), indicating the earliest forms of life were likely **anaerobic** and didn't require abundant free oxygen to live.

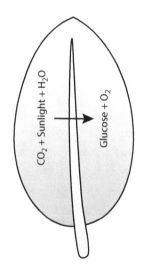

CO_2 + Sunlight + H_2O

Glucose + O_2

Figure 6.14: During photosynthesis, sunlight enables carbon dioxide and water to react and form nutrients and oxygen.

Although photosynthesis may have begun at least 3.5 billion years ago, free oxygen built up slowly in the atmosphere. Rocks from about 2.5–2 billion years ago provide evidence of the increasing concentrations of atmospheric and oceanic free oxygen, although as recently as the start of the Cambrian, free oxygen was likely a mere 10 percent of its current concentration.

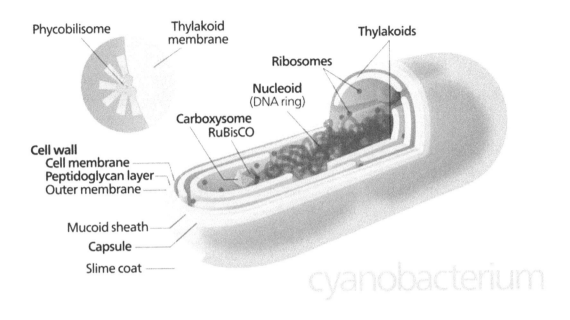

Figure 6.15: Anatomy of a cyanobacterium, a pioneer of photosynthesis.

BANDED IRON FORMATIONS

Banded iron formations (BIFs) are thinly layered rocks that range in age from about 2.5 to 1.8 billion years old and provide a record of the increase of free oxygen in the atmosphere and oceans during that time. The layers are typically a few millimeters to centimeters thick and are composed of alternating beds of silver to gray iron-rich minerals such as hematite (Fe_2O_3) and magnetite (Fe_3O_4) and red, silica-rich chert (Figure 6.17). The difference in color between these layers gives the rock a striped, or banded, appearance. BIFs are the primary source of iron ore.

Figure 6.16: Live stromatolites along Shark Bay, Australia. These are descendants of the oldest types of colonial organisms whose fossils date as far back as 3.5 billion years ago. They have remained relatively unchanged for billions of years.

Figure 6.17: 2.11 Billion-year-old banded iron formations at Jasper Knob, Ishpeming, Upper Peninsula, Michigan.

The majority of BIFs were deposited between 2.5 and 2.0 billion years ago, with a peak at about 2.3 billion years ago, when it is thought that cyanobacteria became multicellular and the rate of photosynthesis increased. Prior to this period, which is called the **Great Oxygenation Event**, the atmosphere and oceans had little free oxygen. Throughout the Archaean, reduced iron continuously accumulated in the anoxic oceans, remaining more or less in a dissolved state. Much of the iron was pumped out from deep-sea hydrothermal vents and dumped into the oceans by rivers. When cyanobacteria began to release free oxygen as a result of photosynthesis, the iron that had built up in the oceans began to oxidize and form little flakes of rust-like minerals that gently rained down to the bottom of the oceans. These tiny flecks of iron oxide accumulated and formed iron-rich layers on the sea bed, especially atop the continental shelves.

BIFs are important because they indicate our atmosphere had become enriched in free oxygen as a result of cyanobacteria. The onset of photosynthesis had a major influence on the chemistry of the oceans and the rocks that were deposited in them. Today, mainly as a result of photosynthesis, free oxygen makes up roughly 20 percent of our atmosphere.

PRECAMBRIAN LIFE

From about 3.8 to 2.1 billion years ago, an amount of time spanning more than a third of the Earth's existence, life on our planet consisted of various forms of bacteria and archaea. These prokaryotes were single-celled and relatively simple in design. Prokaryotes reproduce asexually through binary fission; as a result, two genetically identical cells are produced, and only very limited amounts of genetic information was able to transfer between cells through the process of *conjugation*. Therefore, evolution of these early organisms was stunted by the limited transfer of genes. It wasn't until eukaryotic cells evolved that the process of evolution sped up. Eukaryotic cells differ from prokaryotes in the following ways:

1. Eukaryotic cells contain **organelles**, which are specialized membrane-bound subunits of cells that serve specific functions. For instance, **mitochondria**, found in most eukaryotic cells, produces ATP through the oxidation of glucose and is the source of chemical energy in the cell. The **Golgi apparatus**, on the other hand, takes simple molecules and combines them into more complex molecules for use in the cell or elsewhere. In plants, the **chloroplast** is an organelle that contains the chlorophyll and produces energy for the cell.

2. Eukaryotes, organisms that are composed of eukaryotic cells, reproduce sexually, and offspring are the result of shared genetics from both parents. This leads to an increase in genetic variability and thus greater diversity of a species through time.

3. Eukaryotes are aerobic, meaning they require free oxygen to survive. It is logical, then, that eukaryotic organisms did not come about until after the Great Oxygenation Event.

An interesting thing about the organelles in eukaryotic cells is that some of them, such as mitochondria and chloroplasts, contain their own distinct genetic code, suggesting they had been separate organisms in the past that had "moved into" the host cell. This theory, called the **endosymbiosis theory** (Figure 6.18), posits that eukaryotic cells arose when prokaryotes either ingested other cells or were invaded parasitically, and instead of the enclosed cell dying, it entered into a symbiotic relationship with the host cell. This is one explanation for the origins of mitochondria and chloroplasts, the energy-producing organelles of eukaryotic cells. In this

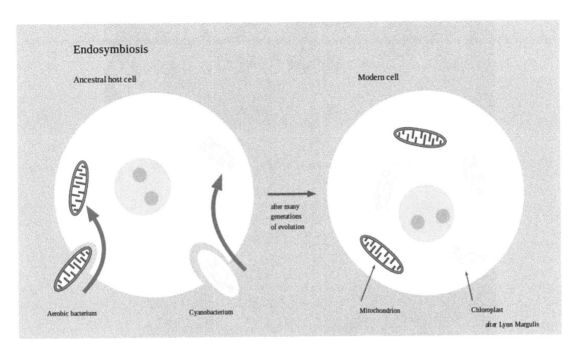

Figure 6.18: Simple diagram showing how endosymbiosis may have led to more complex cells. The origins of mitochondria and chloroplasts may lie among the prokaryotes.

relationship, the host cell provides shelter to the mitochondria or chloroplasts, which produce energy for the cell to use in return.

The earliest possible eukaryotes in the fossil record are about 2.1 billion years old and are called acritarchs (Figure 6.18). Acritarchs, which are thought to be early, spherical, single-celled algal plankton, became the most common type of fossil found in Proterozoic rocks.

MULTICELLULARITY

The development of multicellular organisms is especially important because morphology-driven evolution, that is, the natural selection of specific body shapes, can only occur once organisms develop complex and variable designs. Multicellularity also permits organisms to grow and develop specialized cells that can replace themselves, allowing healing and extended lifespans. The fossil record has yet to indicate the transition to multi-celled life, but it likely occurred within colonies of single-celled organisms that became dependent on one another for survival.

Subject to debate, the oldest multicellular organism may be *Grypania spiralis* (Figure 6.19),

Figure 6.19: Acritarchs, spherical single-celled algal plankton, are a common Proterozoic fossil.

Figure 6.20: Possible tracks of early Ediacaran, bilaterally-symmetrical multicellular organisms within a bacterial mat.

Figure 6.21: This is what today's world would look like if another "snowball earth" period took place. We are currently headed in the opposite direction of this.

found in the 2.1-billion-year-old BIFs of upper Michigan. *Grypania* was possibly a coiled form of algae and measured about 10 centimeters long and about 2 millimeters wide, like a piece of spaghetti. It is also the oldest known **macrofossil**, being visible to the naked eye.

Rocks from Somerset Island, Canada, from about 1.2 billion years ago contain fossils of *Bangiomorpha pubescens*, the oldest taxonomically resolved multicellular eukaryote thus far. Its name alludes to the observation that it is the earliest organism known to sexually reproduce. *Bangiomorpha* took the shape of microscopic filaments of red algae that may have attached themselves to rocks on the seafloor and reproduced via spores.

By the Mesoproterozoic (1.6–1 billion years ago), life had diversified into three kingdoms: archaea, bacteria, and protista (algae). The next kingdom to show up in the fossil record was the animal kingdom, although early fossils are rare and it is difficult to draw the line exactly when animals begin. The earliest fossils that *resemble* animals are worm- and jellyfish-like organisms found in rocks about 700 to 900 million years old. However, the earliest *accepted* animal fossils date to about 585 million years ago and consist of preserved tracks, trace fossils, of what were likely small, centimeter-long slug-like creatures called *bilaterians* (Figure 6.20) that crawled around the floors of shallow oceans in the Late Neoproterozoic. In the intervening period of about 715–635 million years, geologic evidence suggests early life endured at least two catastrophic episodes of glaciation. The **Snowball Earth theory** suggests that the Earth may have been glaciated

Figure 6.22: Rocks from the Ediacara Hills, South Australia.

even at the equator during this time (Figure 6.21). Although fossils of the Neoproterozoic are predominantly from soft-bodied organisms, small fragments of what may be shells and spines are found in these rocks as well.

By 560 million years ago, what may be early forms of worms, jellyfish, and arthropods had evolved, as their fossils are relatively abundant and found in rocks scattered across the globe. Collectively, these organisms are referred to as the **Ediacaran fauna**. These creatures represent the first major diversification event of animals, a time when evolution was rapid and produced a diverse population of organisms in a relatively short period of time. These fossils were first discovered in 1947 by an Australian geologist, R.C. Sprigg, in the Ediacara Hills (Figure 6.22) in South Australia, and have been found on all continents except Antarctica since then. In 2014, two modern specimens of a newly-discovered, strange organism called Dendrogramma were described. They have been compared to the early forms of life such as the Ediacaran Fauna. These mushroom-shaped creatures are similar to jellyfish, but have no stinging cells, making them hard to classify. They may represent a living fossil not unlike some of the earliest macroscopic creatures.

Figure 7.1: A red stalked sea lily (crinoid) (*Proisocrinus ruberrimus*). Crinoids appeared in the mid-Paleozoic and have changed very little since then.

PALEOZOIC LIFE

MARINE ECOSYSTEM TERMS

It is useful at this point to define some common terms used to describe the behavior of marine organisms (Figure 7.2).

Pelagic organisms live in the water column and can be subdivided into two types. **Nektonic** organisms swim through various means of propulsion, such as the fins of fish or the jet propulsion of cephalopods. Some jellyfish can swim by contracting their bells repeatedly. **Planktonic** organisms are passive floaters in the ocean and drift where the currents take them. These include many types of plant (phytoplankton), animal (zooplankton), and a broad, loosely-related group of species called *protists*.

Benthic organisms live on or in the seafloor. Animals and plants that live on the ocean floor are called **epifauna** and **epiflora**. **Mobile** epifaunal animals move around, while **sessile** epifauna and epiflora are attached to rocks or anchored into the mud. **Infauna** are animals that live within and/or burrow through sediments.

CAMBRIAN EXPLOSION

The **Cambrian Explosion** was not an actual explosion, but rather a name for a period of diversification that took place within a short, approximately 10-million-year time frame about 540–530 million years ago. It was not the first diversification event, however; the Neoproterozoic Ediacaran fauna (Figure 7.3) represent an earlier radiation of soft-bodied animals. The Cambrian Explosion resulted in a diversification of skeletal and shelled organisms.

The earliest evidence of shell-like substances being secreted by organisms are calcareous tube-like structures formed by Ediacaran worms. Other Ediacaran fossils such as *Kimberella* also exhibit what may be primitive protective shells. Rocks from the beginning of the Cambrian contain tiny shell fragments from organisms such as the slug-like *Lapworthella* and spherical shells from the mysterious *Archaeooides*. During the Cambrian Explosion, however, much larger animals with skeletons and shells evolved. The reason for

Figure 7.2: A) A coral reef on the north coast of East Timor shows a variety of benthic organisms; B) Reef shark contemplating a nektonic dinner near Sipadan Island, Malaysia.

this is, as always, speculative, but it may have been in part due to a change in the chemistry of the oceans, allowing mineral extraction and secretion to take place more easily.

From the Cambrian Explosion emerged a plethora of new examples of crustaceans, echinoderms, sponges, mollusks, and chordates (including vertebrates), the descendants of which are still around today.

The evolution of exoskeletons such as shells was likely catalyzed by numerous environmental forces, but one that was likely to be important was the need for protection from predators and other dangers of the environment. The lowest trophic levels of the marine food webs were, and still are, predominantly photosynthetic plankton, which would have been restricted to the **photic zone**, that is, the upper region of the ocean where light penetrates and photosynthesis could take place. Organisms that consume cyanobacteria and algae would need to adapt to UV radiation from the sun as well as the possibility of being stranded on the shore during low tide. Shells and exoskeletons provide several lines of defense:

1. Protection from predators: Ediacaran fossils include some organisms with calcified patches on their bodies, giving them an evolutionary advantage against the attacks of predators. This trait would have been selected over time so that by the Cambrian, many organisms had some form of exoskeleton.

2. Increase in size: Shells (and skeletons, for that matter) allow an organism to grow in size by providing a rigid framework upon which vulnerable soft tissues are supported. Selective forces such as an abundant food source and sexual competition favor larger organisms.

3. Protection against UV light: Ultraviolet light damages tissues, and an exoskeleton shields the organism from much of this radiation. This is especially useful for organisms living near surface waters or in intertidal zones.

4. Desiccation: In intertidal zones, organisms stranded on land during low tide can be protected from drying out by hiding in a shell until the tide comes back in.

PALEOZOIC INVERTEBRATES

One of the best preserved and well-known Cambrian fossil assemblages is the Burgess Shale (Figure 7.4) of British Columbia, Canada. Discovered in 1909 by Charles D. Walcott, an American paleontologist from the Smithsonian Institution, this crumbling outcrop of **shale** yields a massive and diverse collection of organisms dating to about 510 million years ago. Organisms represented by these fossils are not limited in extent to the

Kimberella - Possibly an early mollusk, Kimberella was bilaterally symmetrical and had what appears to be shell-like covering that helped protect the animal. It likely lived partially buried in the sea floor.

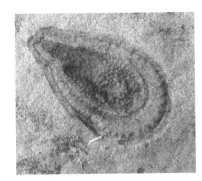

Charnia - Reaching up to a meter in length, Charnia was one of the largest and most widespread of Ediacaran Fauna. Like modern-day sea pens, Charnia likely represents early colonial soft corals.

Tribachidium - This organism is of a yet-unknown affinity. Some scientists speculate this may be a type of cnidarian, although others suggest it is a type of echinoderm. It has three-fold symmetry and a disc-like shape.

Cyclomedusa - A very common Ediacarn organism, Cyclomedusa may have been an early type of sea anemone. It reached diameters of almost a meter across.

Spriggina - Fossils of Spriggina indicate it may have been a segmented worm or possibly an early, soft-bodied arthropod. If the latter is true, this may be an ancestor of Cambrian Trilobites.

EXAMPLES OF EDIACARAN FAUNA

Figure 7.3: Examples of Ediacaran Fauna.

Opabinia - One of the stragner examples of organisms in the Burgess Shale, this creature of unknown affinity had five eyes and a long proboscis that might have helped it dig into mud to search for food.

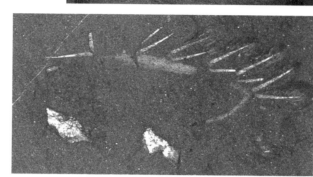

Hallucigenia - As it's name implies, this creature is bizarre in design, making interpretations of it's lifestyle difficult. LIke a modern day velvet worm, this creature was sort of like a cross between a worm and an arthropod.

Ottoia - This carnivorous worm-like organism likely created U-shaped burrows into the mud, and extended it's head into the water to catch prey floating or crawling by. Some fossils of Ottoia indicate they were also cannabilistic, as some fossils include fragments of their brethren in their gut.

Marrella - One of the most common fossils of the Burgess Shale is of Marrella, an arthropod with two sets of spikes that point back from it's head and a couple dozen body segments, each with a pair of legs.

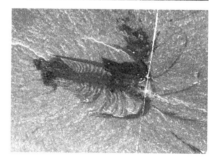

Anomalocaris - A predatory arthropod reaching lengths of 2 meters, Anomalocaris had a ring of "teeth" that closed in like a camera shutter, and two tooth-like prongs that helped push food down it's gullet. It swam gracefully using numerous overlapping fins along the sides of its body.

Figure 7.4: Examples of Burgess Shale Fauna.

Burgess Shale; they can be found in Cambrian rocks throughout the world. However, these particular fossils are some of the best preserved, as they were rapidly buried in fine-grained mud when part of the continental shelf they were living on collapsed into the deep abyss below.

Organisms such as the *Hallucigenia* and *Anomalocaris* are indicative of the diverse group of organisms, many long extinct, that evolved during the Cambrian. This era represents a period during which "experimental" organisms evolved: creatures that tested out new body designs, some successful, and some not. The following pages describe many of these new organisms as they evolved during the Paleozoic and the marine communities they composed (see Figure 7.18).

By the Cambrian, reef systems were teeming with life. They were not the only places in the ocean where organisms existed, but they contained the most diverse and complex ecosystems. Although the first evidence of corals comes from Ediacaran rocks, Cambrian reefs were, according to current understanding, devoid of corals. Early cone-shaped sponges (phylum Porifera) called **archaeocyathids** (Figure 7.5) were abundant in these reefs.

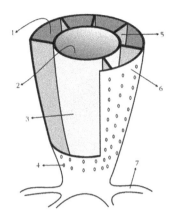

Figure 7.5: Anatomy of an archaeocyathid: 1) Gap (intervallum); 2) Central cavity; 3) Internal wall; 4) Pore (all the walls and septa have pores, not all are represented); 5) Septum; 6) External wall; 7) Rizoid.

Sponges feed by filtering nutrients from water that passes through their porous bodies. Sponge cells, although specialized to perform particular tasks, are more independent of each other than those of other organisms. A sponge can be separated into individual cells that can subsequently reorganize themselves back into a sponge.

Living amongst the sponges were small, inarticulate **brachiopods**. Similar in appearance to bivalves, brachiopods (Figure 7.6) are distinctly different organisms, having a feathery **lophophore** that sticks out and creates a weak current in the water, drawing nutrients in for collection (bivalves have gills). Brachiopods are still around today, but in much smaller numbers than bivalves.

Early brachiopods were inarticulate, meaning they had shells that weren't hinged and able to open. Later in the Paleozoic, brachiopods became predominantly hinged. In these articulate brachiopods, each shell has its own line of symmetry from the hinge to the outer lip of the shell. The two shells, however, are not mirror images of each other. Bivalves, on the other hand, have shells that do typically mirror each other. Each valve does not have its own symmetry, but rather a small "beak" encircled by concentric growth lines.

Crawling amidst the archaeocyathids and brachiopods were *trilobites* (Figure 7.7), distinct and well-known arthropods of the Paleozoic. Trilobites are the most diverse class (over twenty thousand species) of extinct organisms

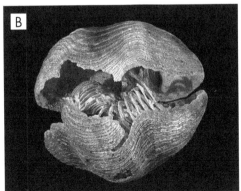

Figure 7.6: A) Example of fossil brachiopods (Paraspirifer bownockeri); B) Well-preserved specimen showing lophophores intact.

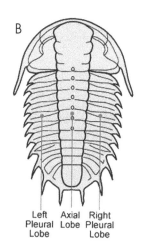
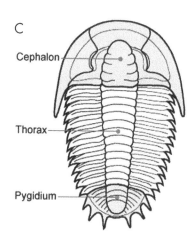

Figure 7.7: A) Redlichiidae trilobite from the Early Cambrian; B) The three lobes of the "tri-lobe'ite"; C) Names of the head area (cephalon); body area (thorax); and tail (pygidium).

known. They were hard-shelled, segmented arthropods that showed up in the Cambrian and lasted until the end of the Paleozoic. Trilobites were benthic, and most were mobile epifaunal creatures that crawled or swam along the seafloor, hunting or scavenging for food as most modern-day crustaceans do. Trilobites were also the first organisms to have eyes that can see shapes and movement; prior to this, eyes were only advanced enough to sense light and shadow. Throughout the Paleozoic, trilobites diversified into a wide array of shapes and sizes, mostly through the modification of the exoskeleton. Some buried themselves in the sand, while others swam above the ocean floor. Their feeding habits also varied between scavenging, grazing, and predation.

The earliest known **cephalopods** evolved in the Middle Cambrian and were similar to today's nautiloids, but were smaller, with elongate, tapering shells. They may have evolved from gastropods in the Early Cambrian that began to use trapped gas in their shells for buoyancy and balance. Most of these cephalopods went extinct at the end of the Cambrian, with the exception of one particular type with a curved shell.

Conodonts (Figure 7.8) remained mysterious microfossils until recently, when the organism thought to grow them was determined to be a primitive jawless vertebrate from the Late Cambrian. Conodonts are the only hard parts preserved from the eel-like organism and likely aided somehow in feeding, similar to the "teeth" of hagfish. These did not have jaws, but rather filter-fed nutrients such as algae and detritus from the ground. They reached lengths of nearly half a meter, although most were several centimeters long.

Cambrian reefs also exhibited various types of early **echinoderms**, although these were primitive and unlike the starfish and sea urchins that typify modern echinoderms. Many other less common invertebrates inhabited Cambrian reefs as well, but soon went extinct.

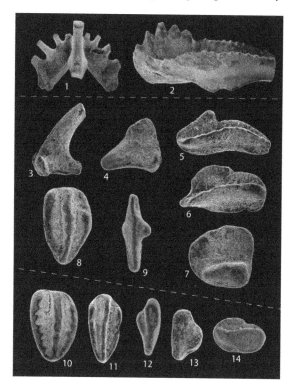

Figure 7.8: Conodonts, teeth of an early vertebrate.

ORDOVICIAN MARINE COMMUNITIES

During the Middle Ordovician, a major marine transgression occurred that inundated the land and increased the extent of shallow seas in which reefs could form. This initiated an important episode of adaptive radiation that heralded a plethora of new types of organisms yet again. With archaeocyathids now extinct, the structural components of the reef were replaced with corals and bryozoans. Early corals may have existed as far back as the Neoproterozoic, but it wasn't until the Ordovician that they became common, mostly in the form of **rugose** and **tabulate corals** (Figure 7.9).

Bryozoans (Figure 7.10), which still exist today, first appeared and became quite numerous during the Ordovician. Most bryozoans were, and are, colonial and form branching structures similar in appearance to coral, but different in the anatomy of their occupants. Other types of bryozoans may instead encrust rocks or even live solitary lives. Unlike coral polyps, bryozoan *zooids* filter-feed using a lophophore, like that of a brachiopod. Bryozoan zooids build protective circular walls composed of chitin (chitin is a hard, organic substance secreted by some marine organisms as a form of exoskeleton) or calcium carbonate, and typically only the lophophore extends out beyond it during feeding.

Due to their rapid evolution and limited time spans of individual species (as little as a million years), **graptolites** (Figure 7.11) make good index fossils for the Early Paleozoic. They first showed up in the Cambrian as benthic, sessile colonies of zooids but became prolific as a mostly planktonic class of organisms in the Ordovician. Many fossil graptolites are of small, planktonic colonies with zooids just on one side, making them look like little hacksaw blades.

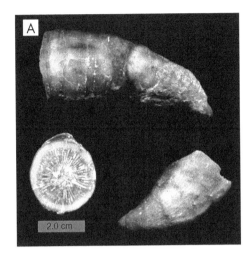

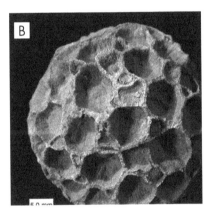

Figure 7.9: A) Rugose coral; note the septa; B) Tabulate coral.

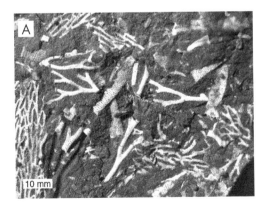

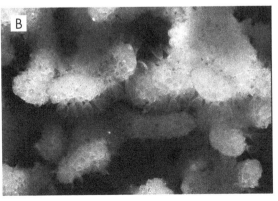

Figure: 7.10: A) Fossil bryozoans from Ordovician oil shales, Estonia; B) modern lacy bryozoan.

Figure 7.11: Fossil graptolites (*Didymograptus murchisoni*).

A new type of echinoderm called a **crinoid** (Figure 7.12) evolved during the Ordovician Period. Also known as "sea lilies," crinoids almost went extinct at the end of the Permian but made it through and are still found in the oceans today. Crinoids consist of a segmented stem, or *stalk*, with a flower-shaped array of filter-feeding arms at the top, forming a *calyx*. The bottom of the stem has a set of root-like appendages called *holdfasts*, that anchor the crinoid to the seafloor. The arms collect nutrients from the water and pass them down to the mouth of the crinoid. The most common fossils of crinoids consist of the de-segmented stems, which consist of small, circular discs that stack up like poker chips. Fossils of the soft calyx are rarer. **Blastoids** (Figure 7.13) were related to crinoids, but instead of

Figure 7.12: A) Early Carboniferous crinoid fossils; B) Disarticulated crinoid stems, the typical form of crinoid fossils; C) Modern crinoids.

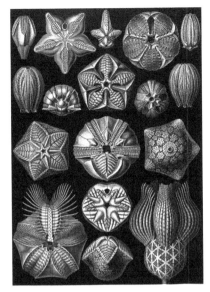

Figure 7.13: Blastoid fossils drawn by Ernst Haeckel.

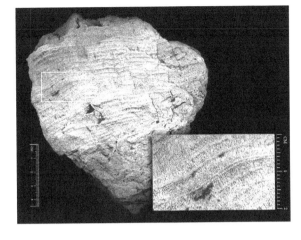

Figure 7.14: A stromatoporoid fossil representing an early sponge.

having an exposed calyx, the feathery nutrient-collecting arms of blastoids were protected by hard, interlocking plates.

Stromatoporoids (Figure 7.14) were mound-shaped sponges with an internal stack of concentric layers similar in structure to that of stromatolites. Some of these mounds reached five meters across, although only the surface of the mound contained the living sponge cells.

Lastly, the Ordovician saw early types of **coralline algae** (Figure 7.15), a type of red algae that extracts calcium carbonate from the seawater and uses it to form protective coatings. These algae are still common today, forming colorful (often pink) mineralized coatings on rocks and other hard substrates. Modern-day

Figure 7.15: Coralline algae on the glass of a reef aquarium.

corallines are useful when determining recent climate change, as the growth rings of these mineral coatings record temperature changes in the seawater.

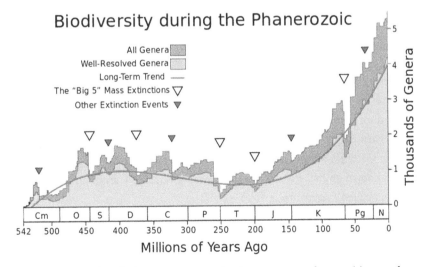

Figure 7.16: The Ordovician mass extinction was one of several known from the fossil record.

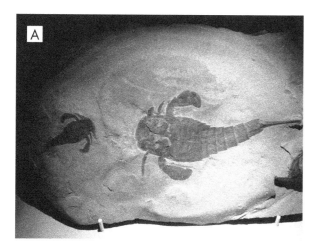

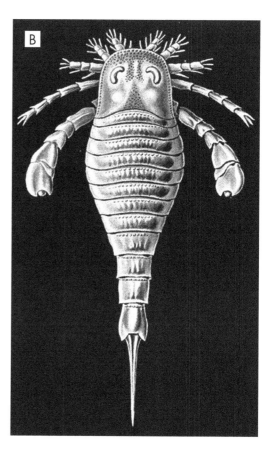

Figure 7.17: A) A eurypterid fossil; B) Sketch by Ernst Haeckel showing a reconstruction of an eurypterid.

Trilobites, cephalopods, and articulate brachiopods continued to evolve and diversify during the Ordovician as well. Although trilobites did not make it past the end of the Paleozoic, brachiopods continue to exist today in limited numbers.

The first major mass extinction event marks the end of the Ordovician Period about 450–440 million years ago. More than 60 percent of all marine invertebrates died, and at least one hundred families went extinct; bryozoans and brachiopods were especially affected. This event took place over several million years and was possibly the result of a severe ice age that took place when Gondwana drifted southward to the pole and began to collect an ice sheet. An ice age such as this would have caused a marine regression, reducing the extent of shallow seas atop the continental shelves and therefore destroying many of the reef environments that had occupied them.

The Silurian period was warmer and a time of recovery for the organisms that had survived the Ordovician mass extinction. Articulate brachiopods diversified and became a dominant organism in Ordovician fossil assemblages, but trilobite species were less numerous than before and never quite recovered. Graptolites, corals, conodonts, gastropods, echinoderms, corallines, stromatoporoids, and bryozoans continued to diversify. Several new organisms evolved during the Silurian as well.

Eurypterids (Figure 7.17) were some of the most dangerous invertebrate predators of the Paleozoic and one of the largest arthropods to ever live, reaching lengths of up to two meters. Referred to as *sea scorpions*, these marine arthropods may be ancestors of today's terrestrial scorpions and possibly other arachnids. Fossils of eurypterids are found in sediments indicative of *brackish* water, a dilute mixture of salty and freshwater typically found in bays and estuaries where rivers empty into the sea. The robustness of their limbs suggests that some eurypterids were capable of leaving the water for short excursions onto land. In fact, the earliest land-based footprint fossils (and therefore the earliest known animal to walk on land) were a possible Cambrian ancestor to eurypterids, a horseshoe crablike organism called *Protichnites*.

Most eurypterids had a long, jointed body and flat, fin-like tail with a spine at the end like that of a scorpion. Long, spine-covered arms may have been used to sift through mud to reach burrowing prey, and the tail was likely held high and used to strike the prey once it was in its grasp. Two large paddle-like appendages were likely used for swimming.

Phylum Cnidaria: This phylum includes corals, jellyfish, and sea anemones. Most cnidarians have stinging, poison-tipped barbs called nematocysts. Corals are composed of many small polyps, which are tiny animals similar to little jellyfish; they have a set of stinging tentacles that surround a mouth opening. The polyp excretes a tiny calcium carbonate theca, or a short, protective tube that the polyp resides in. Individual coral colonies grow by asexual reproduction, allowing the coral head to increase in size with each new generation of polyps. New coral heads arise when the polyps spawn sexually, typically during full moons, and free-floating coral polyps find new homes by attaching to rocks or other coral. Jellyfish and sea anemones are analogous to large coral polyps and either swim or float, as in jellyfish, or attach to a substrate, like sea anemones.

Phylum Echinodermata: This phylum includes organisms with a five-fold radial symmetry, such as sea urchins, sand dollars, starfish, sea cucumbers, and crinoids. The skin of echinoderms is composed of hard, often spiny, plates, and can become vary between soft and rigid depending on signals from the nervous system. These creatures are capable of almost complete regeneration of tissues such as limbs and organs. Echinoderms have no heart, brain, or eyes, but they do have a water-vascular system that allows gas exchange and extracts nutrients from the water. Some echinoderms are carnivorous, such as starfish, while others, such as sea cucumbers, scavenge for detritus on the ocean floor and yet others such as crinoids filter plankton from the water.

coral polyps

starfish

EXAMPLES OF MAJOR INVERTEBRATE GROUPS

Figure 7.18: Major types of invertebrate phyla.

Phylum Molluska: Mollusks are diverse phylum of organisms, but can be subdivided into three main types. Cephalopod mollusks include squid, cuttlefish, the nautilus, and octopuses (yes, that is the correct word for plural octopuses), while Gastropod mollusks include snails and slugs, and bivalves include clams, oysters, and Mussels.

Cephalopods are some of the most complex and intelligent invertebrates. They are able to completely change their shape and color in mere seconds, and use jet propulsion to maneuver through the water, that is, they eject water forcefully from their bodies to move around. While octopuses are soft-bodied, squid and cuttlefish have rigid, but reduced internal shells. The nautilus is the only cephalopod with a spiral-shaped, external shell, and is similar to the extinct, shelled ammonite cephalopods of the paleozoic and mesozoic.

cuttlefish

Gastropods are a diverse group of mollusks that may or may not have a spiral shell. They can found today in a variety of habitats from the sea floor to fresh water and dry land. Those that live in sea water have gills while those in freshwater or on land have lungs. Typically having four sensory tentacles on their head, gastropods have a strong sense of smell and primitive eyesight.

sea snail

Bivalves are molluscs whose bodies are compressed into and protected by two calcium-carbonate shells (valves) hinged together. Most filter feed nutrients from the water through their gills. Some bivalves burrow into sand and mud for extra protection, while others are sessile and attach themselves, or burrow in, to rocks and other hard substrates. Scallops can even swim by repeatedly opening and shutting it's two valves quickly. Some bivalves have a muscular "foot" that helps them maneuver and burrow.

clam

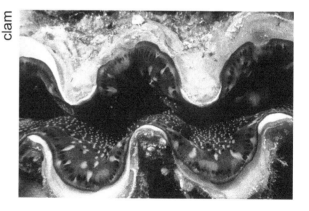

EXAMPLES OF MAJOR INVERTEBRATE GROUPS

Figure 7.18: (*Continued*)

Phylum Arthropoda: Arthropods are arguably the most successful phylum that exists. Nearly three quarters of all living and fossil organisms are arthropods, which include crustaceans, insects, and arachnids. Their name derived from "jointed foot", arthropods have a segmented (jointed) body and a chitin exoskeleton that may be molted periodically. Crustaceans, such as crabs, lobsters, and shrimp, have five or more pairs of legs (decapods) and calcium carbonate inter-grown with their chitin exoskeleton (chitin is a hard, organic substance secreted by some marine organisms as a form of exoskeleton). Arachnids, such as spiders, scorpions, ticks, and mites, have eight legs and well-developed mouth parts, such as fangs. Insects have six legs and many have developed the ability to fly. Arthropods have well-developed senses, including antennae and complex, sensitive eyes.

spider

crab

dragonfly

Figure 7.18: (*Continued*)

EXAMPLES OF MAJOR INVERTEBRATE GROUPS

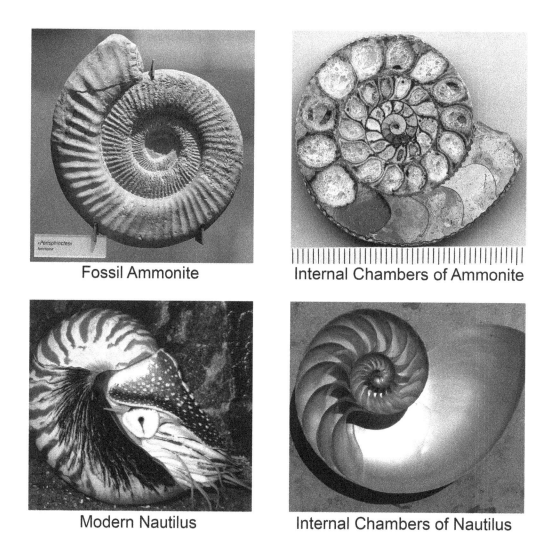

Fossil Ammonite

Internal Chambers of Ammonite

Modern Nautilus

Internal Chambers of Nautilus

Figure 7.19: Comparison of a nautiloid cephalopod with a Mesozoic ammonite.

By the Devonian period, earlier cephalopods had evolved into **ammonites** (Figure 7.19), which left behind well-known and ubiquitous index fossils of the Paleozoic and Mesozoic. Ammonites had elongate or spiral shells and swam through the ancient seas using jet propulsion. One of the characteristics that makes ammonites excellent index fossils is the set of **suture** marks on each shell. Sutures mark the boundaries where internal walls, called septa, intersect with the outer shell wall. The septa separate the shells into distinct chambers, the outermost holding the organism and the other inner chambers filled with gases for buoyancy. The suture lines in early ammonites were simple in shape, but as time progressed, the suture patterns became more complex, reflecting a progressive folding of the edges of septa walls. The early, simple suture patterns are referred to as *goniatitic* and represent ammonites from the Paleozoic. During the Mesozoic, the septa became more complex.

Many other Paleozoic to Mesozoic cephalopods, including the earliest **nautiloids**, had straight, conical shells and belong to the order Orthocerida. A common fossil, often seen as a light gray, elongate cone-shaped shell in polished black mudstones, is known as *Orthoceras*.

Fusulinids (Figure 7.20) were a type of foraminifera (an amoeboid protist still alive today) with a calcium carbonate test (shell). Fusulinids are thought to have been planktonic as well as benthic, lying in or on the mud of the seafloor and feeding off of bacteria and other small plankton using tiny, thin, needle-like *pseudopods* to collect food.

Another organism to evolve were **shrimp** (Figure 7.21), long-bodied crustaceans that mostly swam in the water as opposed to crawling on the ocean floor. They evolved late in the Devonian from earlier benthic arthropods.

Another mass extinction event took place in the Late Devonian, about 374 million years ago. It likely resulted from a change in sea level from global cooling, although the primary organisms to be affected by this were mostly tropical marine invertebrates, with about 50 percent of genera going extinct and many others coming close, such as trilobites, ammonites, stromatoporoids, and coral.

Following the Late Devonian mass extinction, many marine invertebrates re-diversified yet again, but reefs in the Carboniferous and Permian periods were much more restricted in extent due to the decimation of reef builders such as coral and stromatoporoids. Sea levels rose and fell throughout the Carboniferous, resulting in a decrease of continental shelf areas to build reefs upon. Small patch reefs in the Late Paleozoic were primarily occupied by echinoderms (mainly crinoids), bryozoans, brachiopods, bivalves, worms, fusulinids, eurypterids, trilobites, gastropods, and coralline algae, but they generally lacked corals and large sponges.

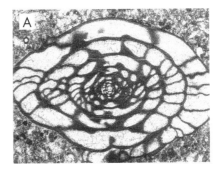

Figure 7.20: A) Cross-section of a fusulinid; B) Intact fusulinid fossils from the Late Carboniferous.

PALEOZOIC INSECTS AND ARACHNIDS

Insects are often avoided in textbooks about prehistoric life mainly because their evolutionary history is vague. Their ancestors were most likely crustaceans, as fossil evidence indicates some of them, such as *eurypterids* and the Cambrian *protichnites*, spent part of their lives on land. This behavior continues today with many crustaceans, such as the coconut crab, which spends its whole life on land and would actually drown if it were submerged in water.

Although insects are commonly thought of as pests, they play an integral role in terrestrial food webs, especially as a pollinator of flowering plants, which make up

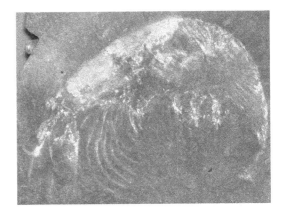

Figure 7.21: Fossil shrimp from the Middle Pennsylvanian.

about 90 percent of all extant plant species today. Some insects even play an economic role: bees produce honey, silk comes from caterpillars, and wild dates are fertilized by wasps.

Despite the sheer number of insects today, the fossil record has not yet provided clear evidence of how and when insects first evolved. It is common for insect fossils to be disarticulated; despite the body segments being hard, they are weakly attached to each other, and it is therefore rare for completely intact insects to fossilize.

The earliest known land-dwelling insects were Silurian millipedes, such as *Pneumodesmus newmani* (discovered by a bus driver and amateur fossil collector). Fossils of *Pneumodesmus* (Figure 7.22) were found in 428-million-year-old siltstones from Scotland.

Figure 7.22: *Pneumodesmus-newmani*, similar to a millipede, is the oldest known insect fossil.

Figure 7.23: Fossil of *Trigonotarbidae* (early spider).

Figure 7.24: Sketch of *Strudiella devonica*, the earliest intact insect fossil.

Centipedes followed not long after, as their molted exoskeletons can be found in the Late Silurian Ludlow Bone Beds from the U.K. Centipedes are similar to millipedes in that they are long, segmented, many-legged arthropods. One of the traits that sets centipedes apart is their venomous claws (called *forpicules*) that likely evolved through the modification of the front two legs.

Early spider-like arachnids (same phylum as insects, but different class) show up as fossils from Late Silurian strata, about 410 million years old, such as the order Trigonotarbida (Figure 7.23), an eight-legged predatory arachnid that differed from true spiders in that it did not contain *spinnerets*, the organs that produce silk. Trigonotarbida (and subsequent true spiders) likely evolved from earlier crustaceans; arachnids and insects are closely related to crustaceans in that they are all considered part of the subphylum Chelicerata, a group of arthropods with distinct feeding pincers or fangs situated close to the mouth. The fossil record indicates that the ability of arachnids to spin webs (and thus be considered "true" spiders) did not appear until about three hundred million years ago during the Carboniferous.

The earliest known fossils of flying insects are fragments (legs, heads, etc.) of *Rhyniognatha hirsti* in a chert formed in a hot spring about four hundred million years ago. A particularly well-preserved jaw of *Rhyniognatha* from the chert indicates it was a flying insect.

The earliest *complete* fossils of an insect come from fossil-rich muddy river sediments dating to about 365 million years ago, in the Late Devonian. *Strudiella devonica* (Figure 7.24) was a small (less than one centimeter), elongate insect with large eyes, a set of long antennae, and possibly wings, although the latter hasn't shown up in the fossils so far.

The Carboniferous was warm, and the seas often transgressed, inundating the land and creating massive belts of swamps. Much of today's coal deposits were formed from Carboniferous swamps and contain numerous fossils of insects, many of them winged, such as dragonflies with wingspans exceeding two feet. The first land snails appeared about 350 million years ago after evolving from what were likely freshwater aquatic snails living in Carboniferous swamps.

The Permian witnessed a major diversification of insects, with many new orders evolving, including those of cockroaches, beetles, and a myriad of different types of flies. Many of these flying insects had developed a new trait: the ability to fold their wings in, and in some cases, protect them entirely beneath an elytra (a hard outer wing), seen in cockroaches and beetles (Figure 7.25). Insects with folded wings have an evolutionary advantage in that they can crawl through tighter spaces and burrow underneath objects or even in the ground. Spiders had also grown more robust during this period, likely as a result of the increase in size and diversity of ground-dwelling insects.

During the Carboniferous Period, many insects and arachnids grew to large sizes. Photosynthesis from the large amount of vegetation growing during this period supplied an abundance of oxygen to the atmosphere, perhaps up to 30%. There are two major theories about why this excess oxygen resulted in giant insects:

1. Insect size is somewhat dependent on the oxygen concentration in the air. There are positive links between the amount of oxygen available and the maximum size to which adult insects and arachnids could grow. Most adult insects can regulate the intake of oxygen through their skin. However, too much oxygen may also prove to be a problem.

2. Recent studies show that insect larvae, many of them deposited in water, cannot regulate their oxygen intake and higher levels of oxygen dissolved in water could poison the smaller larvae. Through natural selection, larger larvae were more likely to avoid oxygen poisoning and would therefore grow into larger insects.

Figure 7.25: *Chalcosoma*, a beetle, with spread wings.

Figure 7.26: *Haikouella lanceolata*, an early vertebrate from the Early Cambrian.

PALEOZOIC VERTEBRATES

Vertebrates are organisms with some sort of bony or cartilaginous, segmented spinal column that protects a centralized nerve cord. They also have a brain, eyes, and a mouth at the posterior end of the spine and are bilaterally symmetrical. Vertebrates include fish, amphibians, reptiles, birds, and mammals and are a subphylum of the phylum **Chordata**, which also include organisms with a flexible (non-segmented) *notochord* in place of vertebrae.

The evolutionary history of chordates is unclear. Candidates for the earliest known chordates come mostly from the Lower Cambrian and include *Yunnanozoon lividum* and *Haikouella* (Figure 7.26), both of which were probably eel-like bottom-feeders. The earliest generally accepted chordate and ancestor to fish, *Pikaia* (Figure 7.27), comes from the Burgess Shale and other similarly aged rocks. *Pikaia* was also eel-like in shape and small in size (generally about two inches long). *Pikaia* is considered to be the forerunner of vertebrates.

The earliest vertebrates date to the Cambrian Explosion and include *Myllokunmingia* (Figure 7.28), found in Lower Cambrian rocks from China. *Myllokunmingia* was a small (less than one inch), gilled ancestor to fish, having a cartilaginous skull and spine as well as a single fin running down its back. Another specimen, Haikouichthys, dates to 525 million years ago, and is considered to the first true fish by *some*.

Figure 7.27: *Pikaia*, perhaps one of the earliest known fish.

Figure 7.28: *Myllokunmingia*, a possible ancestor of fish.

Figure 7.29: Early jawless fish (agnatha) called *Arandaspis*.

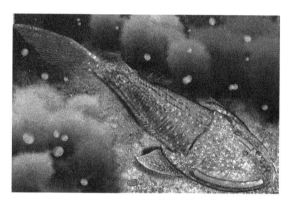

Figure 7.30: Recreation of an ostracoderm fish.

Figure 7.31: Acanthodian Fossil (*Diplacanthus longispinus*).

THE FIRST FISH

Fish are some of the earliest vertebrates and date back to the Late Cambrian; through them came the lineages of all modern vertebrate forms. The earliest fish were jawless and shaped like somewhat flattened eels with underdeveloped fins.

Jawless fish fall under the superclass **agnatha** (Figure 7.29), which simply means "no jaw." They swam around near the shallow ocean bottoms and likely filter-fed nutrients off the mud and rocks. Their swimming style and sets of fins originated in the earlier, eel-like ancestors, but new traits were beginning to develop, such as a more robust body, protective armor made of rigid plates, and the use of gills for respiration only, and not for feeding.

Relatively intact fossils of a cartilaginous jawless fish called *Arandaspis* from the Late Cambrian/Early Ordovician record a fish about six inches long with bony plates protecting its head and tough scales down its body. It had short fins and a tapering tail, making it appear similar to a large tadpole. The armor was likely an adaptation serving to protect them from attacks by cephalopods and eurypterids.

Ostracoderms (Figure 7.30) were "plate-skinned," jawless, bottom-feeding fish that originated in the Late Cambrian. The earliest ostracoderm fossils are scales of *Anatolepis* found in rocks in Australia dating to about 510 million years ago. These scales are made of calcium phosphate, a mineral utilized by vertebrates, as opposed to calcium carbonate, typical of marine invertebrates. *Hemicyclaspis*, another ostracoderm, had developed primitive pectoral fins used in stabilization during swimming. It also had eyes on the top of its head, making it capable of spotting predators more easily.

Without jaws, these early fish were limited to feeding off of detritus on the muddy ocean floor by sucking food into their mouths. The soft gills of jawless fish are supported by long, thin bones called gill arches that act sort of like tent poles. It is likely that the first jaws developed when the front two or three gill arches became hinged and allowed the fish to open its mouth wider to take in more oxygen through the gills. Thus, it may be that the evolution of the jaw was first catalyzed by respiratory needs, and not feeding habits.

The first jaws to ever develop in vertebrates show up in fish called **acanthodians** (Figure 7.31) and **placoderms** (Figure 7.32) that originated in the Lower Silurian and radiated during the Devonian.

Acanthodians were small fish with short, blunt heads, reduced body armor, large eyes, and bony spines supporting several fins, some of them paired up along its underside. Some early acanthodians, such as *Ischnacanthus*, had primitive teeth (made from dermal plates, not socketed) for biting; as time progressed, the layout of the jaw went from one that allowed the teeth to interlock when closed to one that allowed a shearing action, wherein the bottom set of teeth slide up and behind the upper teeth.

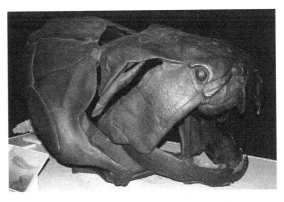

Figure 7.32: Skull of the placoderm *Dunkleosteus*.

Another type of fish, placoderms (meaning "plate-skinned") evolved during the Silurian and became prevalent during the Devonian. Placoderms (Figure 7.32) can be differentiated from Acanthodians due to their overall larger size, greater diversity, greater range (found in sediments ranging from freshwater to the open ocean), and presence of bony, razor-sharp teeth that were part of the skull and not just dermal in origin. They, too, were heavily armored around the head but generally unprotected along the rest of their bodies. Some placoderm fossils are found in sediments indicative of freshwater.

Placoderms were predators that likely hunted mollusks, crustaceans, and other fish. Some reached lengths of up to twelve meters, such as *Dunkleosteus*, which had one of the largest bite forces of any vertebrate, up to eight thousand pounds per square inch (twice that of a Great White shark). *Dunkleosteus* had a jaw lined with extremely sharp, bony plates that could rapidly open, causing a suction of water into its mouth, then snap shut again. It hunted smaller fish and was the top predator in its habitat.

Cartilaginous fish also evolved during the Silurian and radiated in the Devonian. Sharks, skates, and rays are the only types of cartilaginous fish today, and throughout prehistory, they have been relatively restricted in number. The earliest shark known in the fossil record is the Late Devonian *Cladoselache*, which grew to about six feet long. *Cladoselache* had teeth that were cusped rather than serrated and was likely able to grasp prey and swallow them whole.

Lastly, **bony fish** evolved during the Devonian and today are the most diverse and common type of fish. There are two main groups of bony fish based on the anatomy of their fins: **ray-finned fish** (**actinopterygians**) and **lobe-finned fish** (**sarcopterygians**). Ray-finned fish are far more abundant than lobe-finned fish.

Ray-finned fish have thin, flexible fins supported by a set of bony spines that radiate from the joint. The earliest known fossils of ray-finned fish are of *Andreolepis hedei*, which are about 420 million years old and come from rocks throughout eastern Europe. *Andreolepis* possessed primitive teeth that grew from the jaw bones in nested rows, like a shark. Unlike other types of fish, the early ray-finned fish evolved in freshwater habitats such as lagoons, streams, and possibly lakes. It was only later in the Carboniferous did they move into marine environments.

Lobe-finned fish have thicker, muscular fins that contain a set of articulated bones and are jointed to the body in a similar way that a hand is jointed to the wrist. These more robust fins are capable of greater swimming power and speed, and in some cases, walking on land. The earliest lobe-finned fish are from the Late Silurian or Early Devonian, and they quickly diversified by the Mid- to Late Devonian, making them, at the time, more common and diverse than the ray-finned fish. One of the earliest types of lobe-finned fish, **coelacanths** (Figure 7.33), are still around today and are considered to be a "living fossil" because they were thought to be extinct, then were rediscovered in 1938 off the coast of South Africa.

Another important order of lobe-finned fish were **crossopterygians** (Figure 7.34), the ancestors of lungfish as well as tetrapods, including amphibians, reptiles, birds, and mammals. *Rhipidistia*, a member of this group, lived as a predator in freshwater environments, such as rivers and lakes. They were especially adapted to surviving out of

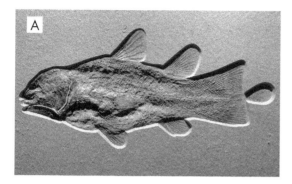

Figure 7.33: A) A fossil Jurassic *coelacanth*; B) A preserved modern *coelacanth* specimen.

the water, as the skeletal structure of their limbs is homologous to that of tetrapods (four-limbed vertebrates), and they had primitive lungs.

Lungfish, an extant group of freshwater lobe-finned fish that evolved not long after the crossopterygians, are modern analogs of the earliest fish to adapt to a semi-aquatic lifestyle. Most nektonic fish have a swim bladder, an air-filled organ that helps with buoyancy. In lungfish, this swim bladder has evolved to be able to absorb oxygen, an adaptation that enables them to survive periods of time out of the water. When in water, however, lungfish breathe through gills.

It is no surprise that freshwater, lobe-finned fish are the most likely candidate for the missing link between fish and amphibians. Like lungfish, the crossopterygians had to adapt to the environmental pressures that accompany a freshwater lifestyle. Several obstacles associated with living in freshwater are provided below and attempt to help explain the theory that links fins to ribs.

1. Plants: Freshwater environments are especially prone to being choked with plants and plant debris, especially within swamps and along rivers. Fish living in shallow bodies of water such as these would be aided by robust, muscular fins, as they would be better able to squeeze and push their way through underwater thickets of vegetation.

2. Periodic Drying: Lungfish developed a secondary breathing organ besides the gills so they could survive when the shallow bodies of water in which they resided periodically dried up or became anoxic. When forced to breathe air, lungfish tend to burrow into the mud for moisture and wait (sometimes months) until the water comes back. Daily tidal fluctuations can have the same adaptive outcome.

3. Gravity: Even a fish with delicate fins can easily swim in open water, but it takes especially strong fins to crawl through the mud. Crossopterygians already possessed the strong backbone and robust fins required for shallow aquatic and terrestrial locomotion.

In fact, the fins of crossopterygians and the limbs of tetrapods share, more or less, the same array of bones. The similarities with the limbs of amphibians are especially noteworthy.

Figure 7.34: Illustration of a typical crossopterygian.

Panderichthys (Figure 7.35) is a Devonian lobe-finned fish fossil found in 380-million-year-old rocks from Latvia and is considered by many scientists to be one of the direct ancestors to amphibians and all other tetrapods. Its head was shaped more like an amphibian than a fish, and its fins had primitive, non-jointed "fingers" at the end of a set of thicker bones, showing an early version of an arm with a hand at the end. *Panderichthys* lived in shallow freshwater estuary-type environments.

Figure 7.35: Illustration of *Panderichthys*.

Because we have strict definitions of what fish and amphibians are, it becomes difficult to classify, in a taxonomic sense, the intermediate forms between the two. Consequently, the species that blur the line between fish and four-legged organisms, referred to collectively as **tetrapods**, are still considered to be fish; in this discussion we will call these intermediate forms "**fishapods**," an informal name adopted by many scientists to refer to fish with tetrapod-like traits. Until recently, the fossil record hadn't provided many of these transitional forms. In 2004, however, a 375-million-year-old fishapod with nearly equal amounts of fish-like and tetrapod-like traits was found in Northern Canada, and it helped fill in this gap.

Tiktaalik roseae (Figure 7.36) reached lengths of up to nine feet, had gills and lungs, scales, a more robust rib cage (upon which muscles involved in walking and breathing could attach), a wide skull with eyes on the top of its head, a flexible neck, and equally strong back and front fins with wrists and finger bones. It lived over ten million years earlier than the first true tetrapods, the amphibians. New data published in 2014 provided more information about the hind fins of *Tiktaalik*; they were surprisingly robust, indicating that strong hind limbs had evolved prior to the first true land-based tetrapods.

Figure 7.36: Fossil of *Tiktaalik* in the Field Museum, Chicago.

Despite the muscular, limb-like fins, *Tiktaalik* spent most of its time in the water eating small fish and possibly insects, and likely made only brief excursions onto land. It may have preferred this habitat to more open waters, where larger predatory fish resided.

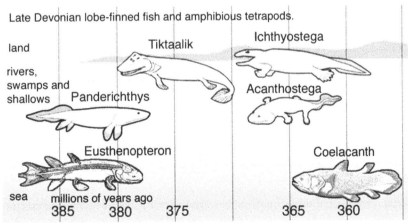

Figure 7.37: Figure showing *Tiktaalik*, with similarities to both lobe-finned fish and early tetrapods, like *Ichthyostega*.

Figure 7.38: Model of *Acanthostega* at the State Museum of Natural History in Stuttgart, Germany.

Figure 7.39: Model of *Ichthyostega* at State Museum of Natural History in Stuttgart, Germany.

As mentioned earlier, it is difficult to draw the line between fish and the first true amphibians (Figure 7.37). Rocks from about 365–360 milllion years ago contain fossils of *Acanthostega* (Figure 7.38), one of the earliest tetrapods and arguably an early primitive amphibian. *Acanthostega* is considered a labyrinthodont, a subclass of amphibian affinity named for its complexly folded tooth surfaces. *Acanthostega* was a little more evolved toward the amphibian end of the spectrum than *Tiktaalik* and was fully capable of walking on land. Despite this, *Acanthostega* still preferred the water.

Acanthostega had limbs with webbed fingers but lacked adequate wrists and ankles. It probably used its limbs primarily for navigating through swampy thickets of vegetation. Besides internal, fish-like gills, it had lungs, but the rib cage was not wide enough to allow it to breathe easily, nor was it robust enough to provide the area on which large muscles used for walking could attach. Its spine was too weak for purely land-based living. Its pelvis and shoulders, however, were strong enough to allow walking on land. The teeth of *Acanthostega* are more like those of animals that reach, bite and grasp prey, and not like those of fish, which tend to use suction to pull food toward their mouths. Thus, *Acanthostega* was likely a predator that lived in swampy areas and used its limbs mostly for navigation in vegetation-choked water.

A later tetrapod that is very similar to an amphibian is *Ichthyostega* (Figure 7.39). Fossils of *Ichthyostega* indicate it too spent much of its time in the water. Its rib cage, pelvis, and shoulder region, however, were stronger than those of *Acanthostega*, indicating it was better at moving around on dry land, and it likely relied on its lungs more than its gills.

The arrangement of bones in its forelimbs and the shape of its vertebrae indicate *Ichthyostega* moved by extending its front half forward, then contracting its abdomen to bring up the hind quarters, similar to how seals walk. It is yet unclear whether or not *Ichthyostega* and *Acanthostega* were contemporaries of each other.

Whether or not *Ichthyostega* was an amphibian is contentious and comes down to where the line is drawn between fishapods and amphibians. The fossil record makes it clear, however, that by the Late Carboniferous, amphibians had radiated and evolved into many types of forms (most of which are unlike today's amphibians). The largest of these were the labyrinthodonts, which reached over two meters in length. Evidence such as jaw structure and tooth wear indicate labyrinthodonts dwelled in swamps and, besides eating plants, hunted fish, insects and smaller amphibians.

Other Late Carboniferous amphibians include *Eogyrinus*, a fifteen-foot-long, eel-shaped amphibian with short legs and a long tail that hunted fish in the swamps of western Europe. The Early Permian *Eryops* (Figure 7.40) occupied swamps of Europe and North America, but unlike *Eogyrinus,* it was shorter, more robust, and had large jaws with sharp teeth.

Figure 7.40: Illustration of *Eryops*.

Figure 7.41: Illustration of *Diplocaulus magnicornis*, an Early Permian lepospondyl.

By the end of the Permian, labyrinthodont amphibians had branched off into three broad types. Lepospondyls were generally small and had moist skin like modern amphibians. Many lepospondyls looked very different from modern amphibians, although some resembled snakes and salamanders. One of the larger examples of lepospondyls (and the earliest) was the mostly aquatic *Diplocaulus* (Figure 7.41), which reached lengths of three feet. *Diplocaulus* had a strange, boomerang-shaped skull and a salamander-like body. Temnospondyls, on the other hand, generally spent more time on land, although the degrees of aquatic lifestyles among them vary quite a bit. They were larger, with more robust bodies and bigger heads.

The third group of amphibians, reptiliomorphs, resembled reptiles most of all, and for some time, many of these species had been misclassified as early reptiles. Reptiliomorphs had long bodies and short limbs, desiccation-resistant skin, and a generally newt- to crocodile-shaped body. They are likely the group that eventually evolved into reptiles.

EVOLUTION OF REPTILES

Reptiles differ from amphibians in many important ways, some of which are listed below:

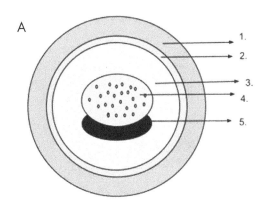

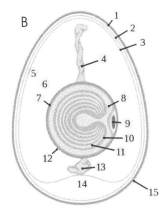

Figure 7.42: A) An amphibian egg: 1) jelly capsule; 2) vitelline membrane; 3) perivitelline fluid; 4) yolk plug; 5) embryo; B) An amniotic egg, in this case that of a chicken, is much more complex: 1) shell; 2) outer membrane; 3) inner membrane; 4) chalaza; 5) exterior albumen; 6) middle albumen; 7) vitelline membrane; 8) nucleus of pander; 9) germinal disk; 10) yellow yolk; 11) white yolk; 12) internal albumen; 13) chalaza; 14) air cell; 15) cuticula.

Figure 7.43: A) *Aneides lugubris*, an amphibian salamander; B) *Lacerta agilis*, a sand lizard.

1. Reptiles have a more advanced egg: Amphibians, like fish, lay gelatinous eggs that must remain in water so they don't dry out and shrivel up (Figure 7.42A). When amphibians hatch, they are still in a larval stage (tadpoles, for example) and must remain in the water until they reach adulthood. Unlike reptiles, amphibians have gills while in the larval stage but lungs during adulthood. Reptiles, however, are **amniotes** (so are mammals and birds), which are organisms that grow an egg containing an amnion; a liquid-filled sac that includes its own food source (yolk), and waste sac (allantois) (Figure 7.42B). These allow reptiles to develop further while still in the egg and hatch as a miniature adult. Furthermore, the "shell" of the amniotic egg is tough, allowing the eggs to be laid on land without drying out. It was this new, advanced egg that allowed reptiles to colonize dry land.

2. Reptiles have tougher skin: Reptiles have skin that is better equipped to retain moisture even when water is not around. Amphibians have a gas-permeable skin that helps them breathe more efficiently while out of the water (Figure 7.43). A drawback of this adaptation is that they must live in or near water so their skin doesn't dry out. Reptiles also have tear glands and a more complete covering of scales than the early labyrinthodont amphibians they evolved from, protecting them further from drying out. Not only are their scales stronger, but they also have sharp nails for defense and traction.

3. Reptiles have a different skull structure: Reptiles have a simpler bone structure along the back of their heads, and many types of reptiles, such as snakes, have a greater degree of cranial kinesis, the ability of the skull to flex along joints between plates of bone. Not all amphibians have teeth, and those that do have them on the upper jaw, toward the front of the mouth, and use them for grasping prey, which they tend to swallow whole. Reptiles have teeth on both the upper and lower jaw; these teeth are replaceable and aid in biting and tearing plants, insects, and meat.

4. Reptiles do not have gills: While amphibians have three ways of breathing (gills, lungs, and permeable skin), reptiles only have lungs, but they can breathe much more efficiently through them than amphibians can.

5. Reptiles (except for those that lack limbs) are better designed to walk on land: Likely an adaptation that allowed them to escape predators by scurrying, reptiles developed a more efficient gait for walking around. Generally, and with many exceptions, amphibians sprawl their limbs a little more than reptiles do. The slightly more erect posture of reptiles allows for more efficient locomotion, which is an important

Figure 7.44: Illustration of *Westlothiana*.

adaptation when traveling primarily on dry surfaces.

It was these adaptations, and many more, that allowed vertebrates to radiate across dry land, better equipped to overcome the hurdles of desiccation and gravity.

Another blurred line exists between amphibians and the first reptiles. Some paleontologists recognize the 338-million-year-old *Westlothiana* (Figure 7.44)

Figure 7.45: Illustration of *Hylonomus*

as the earliest known reptile, although it is also considered by some to be a reptiliomorph. *Westlothiana* shared characteristics of both labyrinthodonts and reptiles. It was a small, possibly burrowing organism with a long, slender body and short limbs. A near contemporary of *Westlothiana*, *Casineria* was possibly the first amniote and the earliest purely terrestrial vertebrate. *Casineria* had clawed, slender legs and a relatively upright posture.

The earliest agreed-upon reptile is *Hylonomus lyelli* (named by Charles Lyell), whose bones are found in 312-million-year-old sand-filled casts of what used to be rotted out, hollow tree stumps and logs (Figure 7.45). Appearing like a small, modern lizard, *Hylonomus* had small teeth and was likely an insectivore. Fossil footprints of

what appears to be *Hylonomus*, including impressions of claws and scales, are found in even older rocks (about 315 million years old) from the same location. *Hylonomus* and other early reptilian contemporaries are collectively called **protorothyrids (stem reptiles)** (Figure 7.46). These diversified about three hundred million years ago when the Earth warmed and many tropical rainforests, previously dominated by laby-rinthodonts, were replaced by drier, more arid habitats.

Protorothyrids are the common ancestor to mam-mals, dinosaurs, and birds. Those that evolved into mammals first transitioned through fin-backed rep-tiles called **pelycosaurs**, which showed up in the Late Carboniferous but went extinct during the Permian Extinction about 248 million years ago. Their descen-dants, however, progressed into mammals eventually.

Figure 7.46: Illustration of a protorothyrid (stem reptile).

Pelycosaurs were a diverse group of reptiles and were the dominant "reptile" group of the Permian. Technically, they fall under the category of **synapsids** (and the first of their kind), which are egg-producing tetrapods (including mammals today), and differ from reptiles in that they have one hole each on the sides of their skulls to allow for the attachment of jaw muscles. True reptiles, on the other hand, have two holes on each side. Some scientists believe that pelycosaurs were also endothermic (warm-blooded) but still needed help to maintain a constant body temperature. Their "sails" were made of a series of spines that supported a webbing of skin, which may have been filled with blood vessels and aided in temperature regulation. For instance, facing the sail toward the sun could have warmed the blood, whereas facing it toward a breeze could allow for cooling.

A popular pelycosaur fossil is that of *Dimetrodon* (Figure 7.47), which appears like a dinosaur but wasn't (the earliest dinosaurs came about in the Triassic, about forty million years later). It walked on all fours, had a blunt, thick skull, and reached up to fifteen feet long. It likely fed on smaller reptiles and amphibians. Although related to the direct ancestors of mammals, it is not considered to be one itself.

Figure 7.47: A) Skeleton of *Dimetrodon*, a fin-backed pelycosaur reptile and ancestor of mammals; B) Illustration of *Dimetrodon*.

Nevertheless, the direct ancestors of mammals are thought to be carnivorous pelycosaurs, which by the Late Permian began to display even more mammal-like traits besides the synapsid skull structure. The evolution of mammals, including their evolutionary roots in the Permian, will be discussed in the next chapter.

PLANTS

Although the metabolic process of photosynthesis evolved around 3.5 billion years ago, the first true plants date much later to the Early Paleozoic. It is theorized that plants evolved when cyanobacteria entered into an endosymbiotic relationship with an eukaryotic host. This may explain why the chloroplasts of plants, the organelle in which photosynthesis takes place, have their own genetic code.

Figure 7.48: Moss, a typical bryophyte.

The earliest plants likely evolved in the oceans and later adapted to freshwater, and by the Ordovician, thin, green films formed on the rocky shores of oceans and lakes.

The first major group of land plants showed up around 470 million years ago in the Ordovician period. Although more numerous and diverse in the Paleozoic, these early plants were most similar to **bryophytes** (Figure 7.48), which still exist today in the forms of mosses as well as liverworts and hornworts. Known as "pioneer plants," bryophytes live in low-lying, moist environments. It is theorized that they evolved from algae.

Mosses are some of the simplest plants. Being *nonvascular*, they don't have true roots, stems, or flowers, and their leaves are tiny, often only one cell thick. Modern bryophytes reproduce via spores that grow into either male or female plants called gametophytes. The male gametophytes produce sperm that fertilize the eggs in the female gametophytes. This produces a sporophyte generation, which then releases new spores that produce new gametophytes, continuing the cycle. These spores will not grow into anything unless they land on a moist surface, which is why bryophytes are limited to wet environments. The earliest plants, although probably not technically true bryophytes, faced similar restrictions to where they could grow. It wasn't until vascular plants evolved did plants spread farther inland.

VASCULAR PLANTS

In order to overcome the hurdles of living on dry land, some plants evolved by producing new types of specialized tissues (Figure 7.49).

Figure 7.49: Example of the various tissues composing a vascular plant.

Cuticle: To prevent desiccation, plants developed a specialized, waxy "skin" called a cuticle that reduces water loss and also keeps harmful bacteria and fungi from entering the tissue.

Lignin and Cellulose: When plants grow, they must overcome gravity, which requires them to produce semi-rigid tissues (such as stems, wood, and the tough outer membranes of cells) to give them strength, flexibility, and overall structural support. The paper this book is printed on is made primarily of cellulose.

Roots: Roots help anchor plants to the substrate they grow on, as well as help them absorb water and nutrients. Without roots, even the plants with strong stems or trunks will easily be blown down by wind or simply tip over due to gravity. The largest and oldest living organism known (estimated at 80,000 years old) is a grove of over

Figure 7.50: A) Illustration of *Cooksonia*; B) Fossil of *Cooksonia*.

forty-seven thousand aspen trees in Utah that share the same massive root system and are therefore one single plant.

Leaves: The majority of modern plants are theorized to have descended from green algae, and the earliest vascular plants that evolved from them lacked leaves altogether. Instead, photosynthesis took place within the stems. Vascular plants later evolved true leaves, which are more efficient in gathering light for photosynthesis.

RHYNIOPHYTES

Figure 7.51: Fossil lycopsid root (*Lepodendron*).

One of the earliest vascular plants were rhyniophytes, which date to rocks from the late Silurian to Early Devonian. Rhyniophytes were small, simple, branching plants with spore-producing organs (sporangia) at the ends of short stalks. They did not have roots, but rather thread-like rhizoids that absorbed water and nutrients from the ground but didn't provide much structural support. A well-known genus of these early land plants (from the late Silurian) is *Cooksonia* (Figure 7.50), some species of which are thought to be rhyniophytes.

LYCOPSIDS

A group of vascular plants to emerge alongside, and possibly from, the rhyniophytes was the lycopsids (Figure 7.51), some of which are still around today as quillworts and several types of mosses. Like rhyniophytes, early lycopsids were small, branching plants with stalked sporangia. By the Late Devonian, however, lycopsids became **arborescent** (tree-like), growing up to ten meters tall, and by the late Carboniferous reached heights of over thirty meters (Figure 7.52).

Figure 7.52: Illustration of a typical Carboniferous swamp environment. (credit: Ludek Pesek, "Carboniferous Swamp," http://www.visualphotos. com/artist/1x20643/ludek_pesek. Copyright © by Science Photo Library. Reprinted with permission.

SPHENOPSIDS

Also a major component of Late Carboniferous swamps, sphenopsids were seedless vascular plants and the ancestors to today's modern horsetails. They were shaped like elongated toilet brushes more so than horsetails (Figure 7.53), however. Genera such as *Calamites* grew up to six meters tall and were concentrated along the edges of swampy bodies of water.

Seedless vascular plants dominated the Carboniferous swamps. Still, these towering forests of lycopsids and other spore-bearing plants were limited to the shores of lakes and streams because they required a moist environment to reproduce. Around the Late Devonian, however, the first vestiges of seeds began to appear, which would eventually allow plants to move onto dry land and dominate the inland landscape.

The precursors to seed-bearing plants were Early to Middle Devonian heterospores, which produced two distinct sizes of spores, the larger of which grew into the female gametophyte and the smaller into the male. By the Late Devonian, these had evolved into the first gymnosperms.

Figure 7.53: Modern horsetail (*Equisetopsida*), appearing today much like early sphenopsids.

GYMNOSPERMS

Gymnosperms (meaning "naked seed") are the earliest form of seed-bearing vascular plants (Figure 7.54). Seeds evolved as a way to protect tiny embryonic plants, giving them a protective housing until they were ready to germinate. Gymnosperms are still around today in the forms of conifers, cycads, and ginkgo. During the time when seedless vascular plants dominated the Carboniferous swamps, early seed-bearing gymnosperms, such

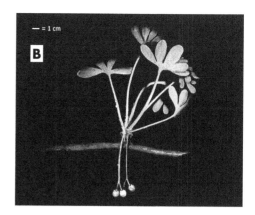

Figure 7.54: Gymnosperms; A) Conifer; B) Ginkgo; C) Cycad

as *Cordaites*, began to form inland forests and grew among other seed-bearing plants such as the ubiquitous *Glossopteris* (Figure 7.55). *Glossopteris* spread across southern Pangaea, and its fossils are found on several continents, serving as evidence of continental drift.

The diverse population of Late Paleozoic seedless plants were more affected by climatic changes in the Permian than the gymnosperms, although the *Cordaites* had also gone extinct by this time. By the start of the Mesozoic, lycopsids and sphenopsids were only present as small, ground-covering plants, and the gymnosperms had become the dominant land plants. The legacy of the Mid- to Late Paleozoic seedless vascular plants is present today in the form of extensive Carboniferous coal belts.

PERMIAN EXTINCTION

The most significant mass extinction event occurred at the end of the Permian about 248 million years ago, wiping out nearly nine out of ten species on the planet. The oceans were the most devastated, with less than five percent of the species surviving the event. On land, trees nearly went away, and about a third of the larger-sized vertebrates followed suit. Although its causes are poorly understood, the Permian Extinction is evident in the fossil record by its devastating effects on the diversity and abundance of life both on land and in the oceans.

There are several theories outlining possible causes of the extinction, and it may be that more than one of these, or even none, were the culprit.

Figure 7.55: Fossil leaves of *Glossopteris*.

1. Volcanism: It is possible that extensive volcanism, which evidently coincided with the extinction event, led to extreme changes in climate. In particular, the Siberian Traps are the remains of one of the most voluminous eruptions of lava in history. Hundreds of cubic miles of lava erupted over tens of thousands of years from broad, flat volcanoes (hypothetically enough to cover the entire planet under twenty feet of molten rock). In some places, the volcanic deposits are over 2.5 miles thick. This amount of volcanism would have also released tremendous amounts of ash and other particles, blocking out the sun, producing highly acidic rain, and leading to global cooling and glaciation. A mass die-off of trees is recorded in rocks from this time, and the strata just above the event boundary are rich in fossils of fungi that would have thrived on rotting wood.

2. Asteroid: There is geological evidence of a seventy-five-mile-wide crater in Australia from rocks of Permo-Triassic age made by an asteroid nearly three miles across. "Shocked" quartz grains, which are caused by a powerful explosion much greater than any nuclear bomb, are found in similarly aged rocks in Antarctica and Australia. This event would have sent a superheated shockwave that ignited forest fires across the globe. The amount of ash and gas from these fires could have had a devastating effect on the biosphere, including acidic rain and global warming.

3. Disturbances in the oceans: The Late Permian was particularly warm, and the southern ice cap on Gondwana may have retreated or completely melted away. This may have led to the slowing down of oceanic currents that would normally mix deep, anoxic water with the surface, allowing aeration of the oceans as a whole. Insufficient oceanic convection could possibly allow anoxic water to build up and suffocate organisms in the shallow water. A different theory states that some physical disturbance in the water, such as a large tsunami, could have stirred up deep, bicarbonate-laden water, causing marine organisms to be poisoned by excess CO_2.

Whatever the cause, the Permian Extinction was a bottleneck through which many organisms barely survived and many more went extinct. Those to never return included fusulinid foraminifera, trilobites, rugose and tabulate corals, blastoids, placoderms, acanthodians, and pelycosaurs.

Figure 8.1: Artist's depiction of a large asteroid heading to Earth, such as the one that marked the end of the Mesozoic.

MESOZOIC LIFE

Before there were big continents, there were little continents. As discussed in Chapter 3, the parts of the lithosphere we call "continents" are thick and buoyant, making them higher in elevation than the ocean surface and therefore composed of dry land.

The early Earth, however, did not initially have continental rocks. The first crust on Earth formed as a frozen rind that covered what was otherwise a semi-molten ball of magma. This early crust was **ultramafic** in composition, as it formed directly from the freezing of surface lava similar in composition to the Earth's mantle. The primitive rock (called komatiite) that composed the early crust was dense, dark, and thin.

As soon as this early crust formed, however, tectonic processes began to reshape it, forming plates that could slide around atop the convecting mantle. When subduction zones initiated, a process called partial melting began, which helped produce the first continental rocks. As the name implies, partial melting involves heating a rock up so that only some of the minerals melt, but not all of them (Figure 8.2).

Since mafic minerals have the highest melting temperatures, they may not fully melt when a rock is heated. Conversely, the first minerals to melt (having the lowest melting temperatures) are felsic. Therefore, by partially melting an igneous rock, a magma will be produced that is more felsic than the original rock from which it came, and a residuum of unmelted mafic minerals are left behind when this new, more felsic magma rises and eventually becomes solid rock.

The felsic rocks produced along the early subduction zones built up and formed small fragments of continental crust, called micro-continents. For a while, these slivers of crust were scattered across the globe, but slowly they began to amalgamate into larger masses along collisional convergent boundaries.

Over time, these smaller fragments of continental crust accreted together into larger and larger continents. Over the next few billion years, these continents combined to form supercontinents, then were ripped back apart by divergence into smaller fragments yet again. Some scientists think at least five to six supercontinents formed and broke apart prior to Pangaea, while others suggest a single, long-lasting supercontinent existed over that span of time and finally broke apart about 600–750 million years ago. Regardless of its span of existence, it is generally agreed that the supercontinent that directly preceded Pangaea was **Rodinia**.

Rodinia existed during a time when life was still very simple; bacterial stromatolites were the most complex forms of life, and they remained in shallow water. Rodinia was an expansive stretch of bare rock weathered by

Figure 8.2: How partial melting creates a more felsic magma; on the left is a mixture of felsic (white), intermediate (gray), and mafic (black) minerals. In the middle panel, the rock is heated so that some of the intermediate minerals melt, and all the felsic minerals melt. In the right panel, the felsic melt has migrated out, leaving behind a concentration of mafic minerals that never got hot enough to melt. The resulting melt (bottom) is more felsic than the original rock.

giant storms born on the global ocean that surrounded it. Between 800 and 700 million years ago, heat that had built up slowly beneath Rodinia began to break it apart, initiating an era of continental divergence and intense volcanic activity. It is known that greenhouse gases are emitted from volcanoes and contribute to the retention of solar heat within the atmosphere. However, the breakup of Rodinia was actually followed by the most intense ice ages ever on Earth, each lasting about ten to fifteen million years.

Geological evidence indicates that during the period of 710–635 million years ago, the Earth had become so cold on several occasions that giant glacial ice sheets spread from both poles and met each other at or near the equator, effectively encasing the planet in a layer of ice a couple miles thick.

One of several possible scenarios leading to *Snowball Earth* is as follows: Much of Rodinia was located in equatorial regions, where the weathering of rocks was high due to large amounts of tropical and subtropical precipitation. Atmospheric CO_2 mixed with water to produce acid rain, which weathered silicate rocks into soil that contained calcium and bicarbonate ions. The soil containing these ions was carried to the oceans and was deposited as layers of calcium carbonate (such as limestone) that trapped CO_2 in rocks at the bottoms of shallow oceans. This period of intense weathering resulted in the removal of huge amounts of CO_2 from the atmosphere. The breakup of Rodinia also allowed what used to be arid land at the center of the continent to be near bodies of water, increasing the rate of precipitation and weathering. This removal of CO_2 from the atmosphere would have reduced the greenhouse effect and catalyzed a period of global cooling. Furthermore, the breakup of Rodinia was accompanied by large amounts of flood basalts, which, when weathered, release large amounts of calcium ions that could be used to create even more limestone and remove even more CO_2 from the atmosphere.

Once atmospheric greenhouse gases are decreased, the global temperatures begin to lower and ice sheets grow. Ice sheets, having a high **albedo** (level of reflectivity), absorb about ten percent of the solar radiation that shines on them (Figure 8.3). Bare rock, on the other hand, may absorb six times as much solar radiation. As ice grows and covers more land, it reflects more and more solar radiation back into space that would otherwise warm the planet. A growing ice sheet, then, creates further cooling, which results in a feedback mechanism that causes the ice sheet to continue growing. If greenhouse gas concentrations are low enough, these ice sheets can achieve a self-perpetuating growth pattern that can spread them across the planet, covering it in ice.

Of course, once the globe became covered in ice, some other feedback mechanism must have kicked in that caused the ice to retreat and the Earth to begin warming again.

Volcanic activity never ceased throughout *Snowball Earth*. For millions of years, their heat as well as greenhouse gas output didn't prevent the ice sheets from growing toward the equator. However, once the rocky crust was covered in ice, weathering decreased dramatically, and therefore so did the formation of carbonate rocks, which normally remove CO_2 from the atmosphere. Without any sufficient mechanisms to trap CO_2 in rocks, it began to collect in the atmosphere instead, leading to a stronger greenhouse effect and a period of warming that began to melt the ice. The same albedo-based feedback mechanism that caused the ice to grow began to work in reverse; as the ice melted, rock was exposed that absorbed much more solar radiation, causing further warming that sped up the melting of the ice sheets.

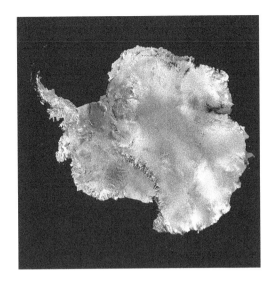

Figure 8.3: Antarctica has a high albedo; nearly 90% of sunlight reflects back off the bright white surface.

Furthermore, the weight of all that ice had depressed the surface of the earth, and when the ice began to melt away, the Earth's surface warped back upward, creating fissures and an increase in volcanism, leading to an even stronger greenhouse effect. In a very short period of time, perhaps on the scale of thousands of years, the ice sheets drew back and the Earth returned to a more normal, warmer climate.

Despite this intense period of glaciation, life in the oceans continued to *just barely* survive. Beneath the thick layers of ice that capped the world's oceans were volcanic fissures and vents associated with divergent boundaries and hot spots. These warm, dark environments were inhospitable to photosynthetic organisms, which require light to survive. Instead, the waters were rich in dissolved minerals and gases, fed by undersea hot springs. Even today, these are common in places where seawater seeps into the crust, warms up to hundreds of degrees, then rises and gushes back out into the otherwise cool, deep ocean water.

Research of these hydrothermal vents led to the discovery, in 1976, of a previously unknown benthic community supported by bacterial primary producers that created food from chemicals, not sunlight (Figure 8.4). In any ecosystem, primary producers are at the "bottom" and play the role of creating organic compounds from those that are inorganic, and in doing so, support the entire hierarchy of organisms in their food web. Although chemosynthetic bacteria had been discovered nearly a century earlier, their essential role in dark, hot, undersea ecosystems became apparent in this new generation of studies.

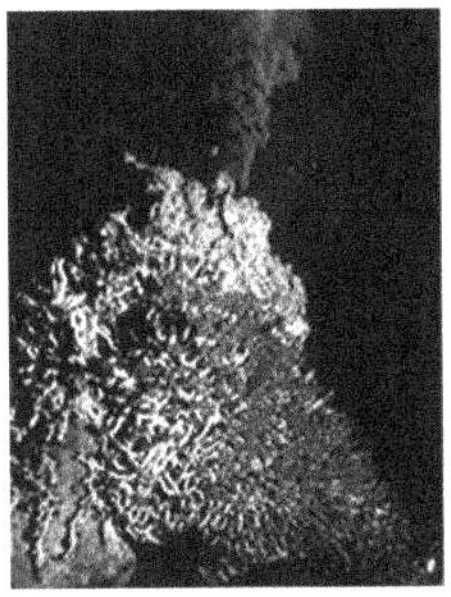

These primary producers are called **extremophiles** because they live in environments previously thought too extreme to support life (Figure 8.5). Through chemosynthesis, these bacteria convert hydrogen sulfide and methane into organic compounds.

Even more recent research has unearthed a plethora of other extremophiles, many of which could have helped life survive Snowball Earth. *Endoliths* live inside fractures in the deep crust or within the minute pore spaces between mineral grains. *Halophiles* can live in very high concentrations of salt. With so much ice, it was likely that the water beneath it was more salty than today. *Piezophiles* can survive the very high pressures found in the deepest parts of the oceans. If Snowball Earth was as extensive as some scientists think, it was the extremophiles that helped enable life to endure it.

Figure 8.4: A community of red-gilled tube worms growing on the Main Endeavour vent along the Juan de Fuca Ridge.

Rodinia broke up twice, actually; it temporarily reassembled about 100 million years later, then finally broke apart in earnest about 550 million years ago.

HISTORY OF RODINIA AND PANGAEA

By the beginning of the Paleozoic, much of the continental landmass we see today had been produced along convergent plate boundaries, although in shapes and arrangements very different from today. The Paleozoic witnessed major mountain-building events along these boundaries as well as several major episodes of sea level rise and fall, called **transgressions** and **regressions**, respectively. During transgressions, the sea level rises and inundates the edges of the continents, creating shallow seas (called **epeiric seas**) atop the continental shelves and pushing the shoreline inward. During regressions, the sea level drops back down, causing the shorelines to move back outward and, in some cases, the shelves to emerge from the water and form wide platforms of exposed land sloping gently toward the oceans. BIFs, for example, formed underwater on continental shelves during marine transgressions.

The breakup of Rodinia created six new continents, all composed of land that would eventually be stitched back together (Pangaea) and ripped apart again into even newer shapes during the Mesozoic. The six Paleozoic continents were called:

1. **Baltica**—What would become most of Russia and Northern Europe

2. **China**—What would become China and Southeast Asia

3. **Gondwana**—What would become Australia, Africa, India, Antarctica, Florida, Southern Europe, Madagascar, and parts of the Middle East

4. **Kazakhstania**—What would become, not surprisingly, Kazakhstan

5. **Laurentia**—What would become North America, Greenland, and parts of Ireland and Scotland

6. **Siberia**—What would become part of Russia, as well as the part of Asia that lies between Kazakhstan and Mongolia

During the Cambrian, these continents were scattered near the equator, and the polar regions of the Earth lacked any significant ice. Later, in the Paleozoic, the continents began to disperse more across the globe. Some contained strictly passive margins (those that are not delineated by a plate boundary), while others were bound by tectonically active margins, often producing significant mountain ranges.

The first major mountain-building event seen in North American rocks, for example, was called the Taconic orogeny (Figure 8.6), which was one of several **orogenic** (mountain-building) events that constructed Pangaea by the end of the Paleozoic. It occurred along the southeastern margin of Laurentia and formed a long, somewhat linear mountain range that continued to grow during the stitching together of Pangaea. This Himalaya-sized

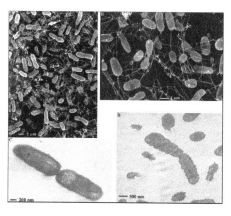

Figure 8.5 Thermophilic bacteria from deep-sea vents. These bacteria metabolize sulfur and hydrogen; A,B: scanning electron micrographs; C,D: transmission electron micrographs.

range can still be seen in fragmented form today as the Appalachian Mountains and other ranges in western Europe, Greenland, and northwestern Africa (the mountain range split into pieces as Pangaea broke up during the Mesozoic).

PANGAEA

Figure 8.6: The Appalachians from space. The snow accents the once-deep-crustal folded rocks that resulted from the Taconic orogeny.

Pangaea (meaning "entire Earth") was almost fully formed by 270 million years ago but began to break apart by about 200 million years ago. During its short existence, many new types of organisms came about, including mammals, *and* the Earth's most devastating mass extinction occurred.

This supercontinent stretched from pole to pole, had a long, continuous western coastline and an eastern coastline indented by a large, shallow body of water called the **Tethys Sea**. Beyond either side was a global ocean called **Panthalassa**, the remnant of which today is the Pacific Ocean.

Pangaea can be separated into two geographical areas; **Laurasia** made up the northern part of the supercontinent and **Gondwana** composed the southern part. With a few exceptions such as India, Laurasia was made up of today's northern continents, and Gondwana of today's southern continents. South America and Africa composed northern (equatorial) Gondwana.

Transgressive and regressive sequences of sedimentary layers dating to the Paleozoic indicate that climate during that time was variable; sea level drops when the Earth cools and rises when the Earth warms up. Another factor influencing sea level is the rate of spreading along divergent boundaries. When spreading is rapid, the seafloor is generally warmer and more thermally expanded, causing sea levels to rise. Prior to the formation of Pangaea, the sea rose and fell repeatedly as the previous generation of continents drifted about. Those that drifted away from the equator saw cooling climates, while those remaining or moving toward the equator were warm. Throughout the Paleozoic, however, climate was warmer globally; by the Early Mesozoic, global temperatures were about 20°C warmer than today and were the result of a strong greenhouse effect (CO_2 concentrations were much higher). The equatorial regions of Pangaea were rainy, with two monsoon-like rainy periods per year. Farther away, the climate was drier and there was only one rainy season. The continental interior of Pangaea was warmer and drier than the coastlines, a pattern observed on today's continents.

MESOZOIC LIFE

Life on Earth barely survived the Permian Extinction. The diversity of life wasn't as affected as the abundance of it. A great number of different species nevertheless survived and went on to evolve into a plethora of new forms during the Mesozoic.

The Mesozoic witnessed a resurgence of life; reptiles evolved into dinosaurs, mammals, and birds, and land plants became dominated by angiosperms, which produce flowers as a mode of reproduction.

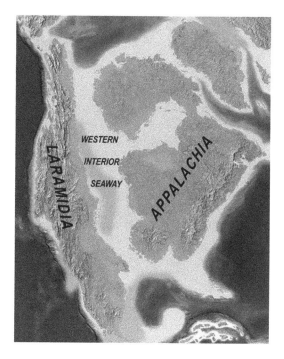

Figure 8.7: The Cretaceous Western Interior Seaway about 75 million years ago. Two temporary subcontinents were formed by the long stretch of shallow sea.

MESOZOIC MARINE REVOLUTION

In marine systems, the Mesozoic was a time when predation increased in reef systems, with new forms of shell-crushing predators and a reciprocal evolution of prey into burrowing and more mobile forms. The Mesozoic also saw an increase in diversity of planktonic forms of life (Figure 8.7).

MESOZOIC INVERTEBRATES

MARINE INVERTEBRATES

Many forms of plankton, arthropods, corals, echinoderms, and other marine invertebrates did not survive the Permian extinction, including trilobites and rugose and tabulate corals. Those that did survive re-diversified later in the Triassic after a brief period of evolutionary quiescence and recovery. Bivalves, cephalopods, and gastropods became dominant marine invertebrates during the Mesozoic.

Ammonites, which evolved in the Paleozoic, became more complex during the Mesozoic. In particular, the septa between each chamber of their shells became more intricately folded, making them excellent guide fossils (Figure 8.8). The simple *goniatitic* suture patterns seen in Paleozoic ammonites were replaced by a *ceratitic* pattern in the Triassic and eventually an *ammonitic* pattern by the Jurassic. While most shells coiled within a plane, some

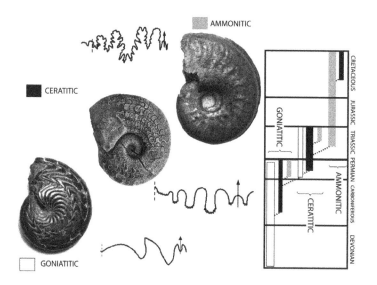

Figure 8.8: Examples of ammonites showing the three main suture patterns and their phylogenetic history.

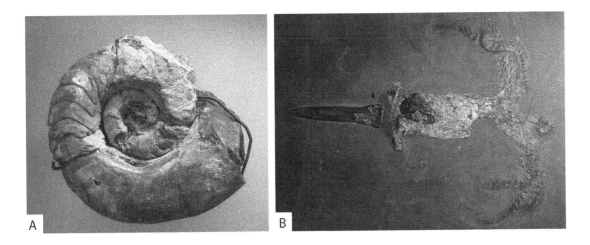

Figure 8.9: A) Fossil nautiloid from the Late Mesozoic; B) A very well-preserved Jurassic fossil of a belemnoid.

shells evolved to coil outwards in one direction, such as the shell of the *Helioceros*, or not coil at all, like that of *Baculites*. Some, like the *Parapuzosia*, had shell diameters over two meters across.

Belemnoids (Figure 8.9), relatives of ammonites, evolved in the Jurassic and lived throughout the rest of the Mesozoic. Like a modern cuttlefish, these cephalopods had an internal bony rod, as opposed to an external shell. This mineralized rod acted as an internal support structure and at one end had several chambers used for buoyancy. Their fossils typically are composed of the internal rod, made of calcite, which resembles a bullet or cigar butt.

Some **arthropods**, such as trilobites and eurypterids, went extinct at the end of the Permian, but many more survived both on land and in the sea. Marine arthropods of the Mesozoic include many familiar to us today. Crustaceans, which have been around since the Paleozoic, diversified during the Mesozoic, especially in the forms and uses of their numerous limbs:

1. **Crabs** (Figure 8.10A) may have evolved as early as the Carboniferous based on a possible fossil *carapace*, the central, upper part of the crab's exoskeleton. Horseshoe crabs also date to the Paleozoic and are still alive today but look dissimilar to the crabs we see in the seafood aisle. By the Jurassic, crabs with long legs and claw-like pincers show up in the fossil record. It is possible that the dominant claw in many crabs and lobsters today and in the fossil record had evolved the ability to break apart the hard shells of bivalves as well as serve as self-protection and competition for mates.

2. **Barnacles** use their legs to collect food particles from the water, while **krill** use their legs to stir up nutrients from layers of mud. **Coconut crabs**, which live solely on land, can use their claws for cracking open their namesake food or for digging burrows in the ground. **Stomatopods** (Figure 8.10B) are mantis-like crustaceans that use their claws to punch holes in the shells of their prey.

Rugose and tabulate corals went extinct at the end of the Permian, but other corals made it through. New types emerged from these survivors, but it wasn't until the mid-Triassic that coral reefs became abundant again. One new and very successful coral type to emerge in the Triassic were the *scleractinians* (Figure 8.11), or stony corals, members of which are still around today as primary reef-builders. While scleractinians appear similar to rugose corals (they both contain septa within the corallite), it is thought they instead may have evolved from a lineage of anemones.

Brachiopods barely survived the extinction and are still recovering, but other bivalves, such as clams, quickly came to dominate reef systems, so much so that by the end of the Mesozoic, they nearly crowded out the corals. By the Cretaceous, **rudist** bivalves became dominant in equatorial reef systems. They are typified by their asymmetric

Figure 8.10: A) *Cycleryon propinquus*, a Jurassic crab; B) *Pseudosculda laevis*, a Late Cretaceous stomatopod.

2.0 cm

Figure 8.11: Jurassic scleractinian coral fossil.

valves; the one on top is rather reduced in size, while the one on the bottom is cone- or tube-shaped and sometimes coiled. They were colonial and appeared superficially like the rugose corals in their growth habit. Cretaceous **inoceramids** were some of the largest bivalves ever, with round, flat, plate-shaped valves up to two meters across. They laid flat on the ocean floors and were often covered in oysters and provided shelter for fish.

Echinoids continued to flourish in the mid- to Late Mesozoic. *Blastoids* had gone extinct at the end of the Permian, but their cousins, the crinoids, continued to inhabit Mesozoic reefs. Crinoids throughout the Mesozoic evolved to be more mobile than sessile, possibly in response to the increased predatory habits of sea urchins, whose bite marks are seen on many crinoid stem fossils from the era.

A common echinoderm in the Mesozoic was the **starfish** (which evolved in the Paleozoic), a mobile, benthic organism with a unique type of vision in which eyes are situated at the ends of their arms. They have "tube feet," which consist of many small, finger-like, hydraulically-driven protrusions on their underside that help them move around on the seafloor and even crack open shells and attack corals (Figure 8.12). They are also scavengers, feeding off of dead organisms by crawling on top of them and excreting their stomachs during feeding and digestion, then pulling them back in.

The lacy bryozoans of the Paleozoic nearly became extinct at the Permian Extinction. Some were able to ride out the chaos and re-diversified during the Mid- to Late Mesozoic, becoming more colonial over time for better protection and competition for space in their reef systems.

The Mesozoic Marine Revolution caused an evolutionary diversification of gastropods, especially snails (Figure 8.13). Those with weak, easily broken shells, however, were preyed upon by crustaceans and declined in number. Meanwhile, those producing thicker, more robust and complex shells were able to thrive. In fact, some gastropods developed the ability to modify shell growth in response to particular predators and even prey on other shelled creatures by burrowing through their shelly carapace. Mesozoic worms, also in response to increased predation in reef and other shallow-water systems, evolved to burrow into the mud to escape predators.

Figure 8.12: A) *Pentasteria longispina*, a Late Jurassic starfish; B) tube feet of a modern starfish.

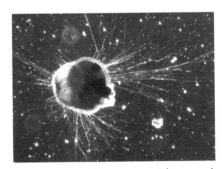

Figure 8.13: *Trochactaeon conicus*, a Late Cretaceous gastropod.

The Mesozoic also witnessed diversification of marine plankton. Whereas fusulinid *foraminifera* went extinct at the end of the Permian, other foraminifera were quite successful in the Mesozoic (Figure 8.14). Foraminifera are protists, which are eukaryotic organisms that are neither plant, animal, nor fungus. Other protists dominant in the Mesozoic were *radiolaria*, which produced intricate shells made of silica.

Another type of plankton common in Mesozoic seas were *coccolithophores*, the primary producers of chalk (Figure 8.15). They were, and still are, single-celled marine plankton that excrete calcium-carbonate coccoliths, which are tiny, round, plate-like shields that surround the cell like a covering of tiny hubcaps. These settle to the seafloor when the coccolithophores die and form layers of calcium carbonate particles, which later turn into carbonate rock. Even today, about 1.5 million tons of calcium carbonate are extracted each year from the ocean waters by these creatures in the form of coccoliths.

Figure 8.14: *Ammonia tepida*, a modern foraminifera.

Diatoms (Figure 8.16) became abundant in the Late Mesozoic and are also still around today. Like coccolithophores, diatoms are marine phytoplankton that are typically unicellular. What makes them unique are their beautiful, complex silica shells they produce in a wide variety of shapes. They play an important role as primary producers in marine food webs. Their delicate silica shells are used today as abrasives and even deterrents to insects such as bed bugs, who don't like to cross the broken glassy fragments scattered strategically around bedposts and box-springs.

Dinoflagellates, also important primary producers, are single-celled protists that are most commonly noted as the reason behind "red tides," where they are so populous they color the seas red. They emit a neurotoxin that often kills other marine life and makes humans sick when they eat infected fish, clams, and crustaceans. Unlike

Figure 8.15: *Coccolithus pelagicus*, a modern coccolithophore.

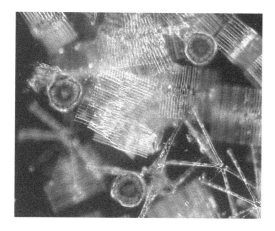

Figure 8.16: Dinoflagellates (small brown spheres), hanging out with some diatoms (clear, delicate structures) in a freshwater lake in Lake Chuzenji, Nikko, Tochigi Prefecture, Japan.

Figure 8.17: Fossil locust, related to grasshoppers, from the Cretaceous.

many other plankton, they are free-swimming, using flagella to move around in the water. Their "armor" consists most typically of a layer of cellulose instead of mineral tests. They likely evolved in the Precambrian, possibly related to acritarchs, but didn't become common in the fossil record until the Late Triassic.

TERRESTRIAL INVERTEBRATES

As discussed in Chapter 8, insects evolved in the Paleozoic from what may have been crustaceans. The Permian witnessed a great diversification of insects. By the end of the Permian, insects had already evolved into many types of cockroaches, beetles, snails, slugs, and flies. In addition, spiders and scorpions, representing members of the arachnids, as well as centipedes and millipedes had evolved into multiple forms. About a third of insect orders did not make it across the extinction event, and those that survived into the Triassic and Jurassic periods mostly went extinct throughout the Mesozoic and into the Cenozoic.

The Mesozoic was a time when insects with foldable wings prevailed (Figure 8.17). Insects that date to the Triassic include the *grasshopper* and the *wasp*. Insects that evolved in the Jurassic include *earwigs, modern flies, caddisflies* (which resemble small, skinny moths), *termites*, and *bees*. About 16 percent of Cretaceous insects are represented by modern descendants, including mosquitos and ants, the latter possibly evolving from earlier wasps. Although the arachnid fossil record from the Mesozoic is sparse, most modern arachnids had evolved by this time and resembled those we see (and often fear) today.

MESOZOIC VERTEBRATES

MARINE FISH

Mesozoic marine environments were dominated by fish, mostly bony, but also a good amount of cartilaginous examples such as sharks. *Teleosts*, types of ray-finned bony fishes, became dominant by the end of the Mesozoic and continue to be the most common and diverse types of fish today.

The largest fish to ever swim the oceans, called *Leedsichthys*, evolved in the Triassic or Jurassic and reached lengths of over fifty feet, despite the theory that it fed only on tiny plankton (Figure 8.18). It is likely that the population boom of plankton during this time led to the increase in size of the fish that fed on them. A fish called *Ischyodus* evolved in the Jurassic and is distinguished by its "bucked teeth," which were actually dental plates likely used to crush shellfish and crustaceans.

Xiphactinus was one of the biggest fish of the Late Mesozoic, reaching lengths of up to twenty feet (Figure 8.19). It had a large, gaping mouth that could swallow prey up to a third of its size, and it is thought

Figure 8.18: An illustration of a Jurassic *Leedsichthys*, the largest fish known in the fossil record.

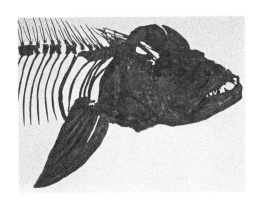

Figure 8.19: Fossil of *Xiphactinus*.

that this fast-swimming fish could even leap from the water to catch unsuspecting seabirds as they flew near, or swam on, the surface waters. *Xiphactinus*, among other large fish, lived in the **Western Interior Seaway**, a shallow inland ocean over North America, during the Late Cretaceous (Figure 8.7).

Known as a "sabre-toothed herring," although not related to herrings, was the Cretaceous *Enchodus* (Figure 8.20). It had large, sabre-like fangs, possibly aiding in the catching of prey such as cephalopods. Although a small- to medium-sized fish, *Enchodus* was still an important predator, as well as prey to larger marine reptiles. *Scapanorhynchus* was a small Cretaceous shark, similar to today's goblin sharks, that likely used electroreceptive sensors on its long, snout-like nose to hunt for fish in deep, light-poor waters.

MARINE REPTILES

Figure 8.20: Fossil of *Enchodus*.

A major radiation of reptiles back into the oceans took place in the Mesozoic era. This trend toward aquatic lifestyles dates to the Triassic, with early examples such as *nothosaurus*, semi-marine reptiles with long necks and tails that likely lived and caught food in the water but spent time on land to sunbathe and lay clutches of eggs (Figure 8.21). They had sharp, spiny teeth and short limbs with paddle-like feet. Like a crocodile, the

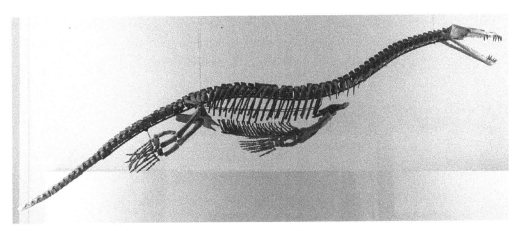

Figure 8.21: Fossil of a nothosaur, a semi-marine reptile.

C

Figure 8.22: Mesozoic marine reptiles: A) ichthyosaur; B) pliosaur; C) elasmosaur.

nothosaur was likely capable of quick bursts of speed in and around bodies of water to catch prey. Another Triassic semi-marine reptile was the *Tanystropheus*. It had an extremely long neck (up to ten feet long) with sharp teeth and a relatively short, yet muscular body. It likely used its small but robust limbs to grip onto rocks or logs along the shorelines and fed on fish and cephalopods by using its long neck like a fishing rod.

Ichthyosaurs (Figure 8.22A), whose name means "fish lizard," were marine reptiles with long, streamlined, fish-shaped bodies, short, powerful "paddles," and crescent-shaped tails. Although they lived in the water their whole lives, they still needed to breathe air, like marine mammals do. *Ichthyosaurs* showed up in the Triassic and reached their peak in the Jurassic.

During the Triassic, many marine reptiles adopted a lifestyle in which they remained in the water. One such group was the *plesiosaurs*, which were typically long-necked, long-bodied reptiles with large, powerful flippers. A subset of these, called *pliosaurs* (Figure 8.22B), grew to be the largest, although they had relatively shorter necks compared to their body lengths. Most plesiosaur fossils are found within shallow-water marine environments, such as those near the coast, but we cannot rule out deeper waters as a habitat at this point. Fossil evidence suggests that at least part of the diet of plesiosaurs consisted of ammonites and belemnites.

Placodonts (Figure 8.23) were Triassic marine reptiles that looked like a cross between a turtle and a walrus, with broad, flattened bodies. They grew to lengths between one and three meters. Placodonts had dense bones and some had armored plates, making them neutrally or negatively buoyant, allowing them to feed easily on bottom-dwelling shelled organisms. Given their relatively small size and lack of flippers, it is possible that placodonts

Figure 8.23: Illustration of *Liopleurodon*.

resided on land part of the time to avoid larger predators such as fish, as well as to lay eggs. They likely would have been slow and sluggish on land and only slightly more agile in the water. Placodonts weren't very successful and went extinct by the end of the Triassic.

In the Mid- to Late Jurassic, a giant predatory marine plesiosaur reptile called *Liopleurodon* (Figure 8.24) rose to the top of the food chain. Reaching lengths of over fifty feet, this predator was an adept hunter, targeting other large fish and reptiles such as *Leedsichthys* and Ichthyosaurs. Its large mouth was full of sharp, dagger-like teeth, and its body was long, having four large flippers helping it to maneuver well in the water. It lived its entire life in the oceans, but, like a whale, had to surface to breathe air.

Figure 8.24: Fossil placodont, an extinct turtle-like semi-marine reptile.

The graceful, Late Cretaceous *Elasmosaurus* (Figure 8.22C) had one of the greatest neck-to-body ratios of all the plesiosaurs. In fact, its neck alone had over seventy vertebrae, making it very flexible and a useful tool while hunting. It had sharp but skinny intermeshing teeth and a rather tiny head, making it likely that its diet consisted of small fish. It is thought that *Elasmosaurus* hunted by swimming beneath schools of fish and raising its long neck into the crowd to feed, somewhat like picking apples from a tree. Gastroliths associated with *Elasmosaurus* fossils suggest it swallowed fish whole, a theory supported by the shape of its teeth, which would have been good at impaling fish but not chewing them.

Mosasaurs were a successful group in the Late Cretaceous but did not survive the extinction event. They reached lengths of over sixty feet and had long, tapering bodies with strong front fins. Their heads were relatively large and contained sharp, conical teeth used to catch fish and cephalopods, among other prey. Mosasaurs likely hunted near surface waters, where they would need to go to breathe. Recent fossil evidence suggests that mosasaurs evolved from smaller, semi-aquatic lizards rather than the plesiosaurs.

TERRESTRIAL AMPHIBIANS

Labyrinthodont amphibians such as lepospondyls and temnospondyls had their most successful period in the Carboniferous and Early Permian swamps. Lepospondyls went extinct during the Permian, and temnospondyls began to decline in numbers into the Mesozoic, in part due to competition from reptiles such as early crocodiles that arose in the Mid-Triassic. By the Cretaceous, they were gone.

Members of the temnospondyl order were the dominant group of amphibians in the Triassic. They ranged in length from a few centimeters to several meters, and their lifestyles varied from semi-aquatic to aquatic. *Metoposaurs*, for example, are common in Triassic and Jurassic swamp deposits in the southwest United States. They had large, broad heads and crocodile-like bodies. The agile *trematosaurs*, on the other hand,

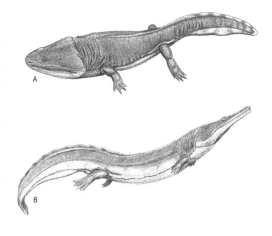

Figure 8.25: A) Metoposaur; B) Trematosaur.

had narrow skulls, elongate bodies, webbed feet, and lived aquatic lifestyles, hunting fish (Figure 8.25).

Figure 8.26: A) *Gerobatrachus*, also known as a "frogamander"; B) A Late Cretaceous fossil of what is likely *Palaeobatrachus gigas*, an early frog.

Figure 8.27: Illustration of a protorothyrid, a "stem reptile" from which mammals, dinosaurs, and many other lineages diverged.

Modern amphibians such as frogs and salamanders are considered members of the *Lissamphibia* clade. The origins of lissamphibians are debated; some scientists place their roots in Permian temnospondyls, while others consider them descendants of lepospondyls. More research needs to be done to better describe the evolution of amphibians during the Late Paleozoic and Mesozoic.

One promising lead is the 290-million-year-old fossil of *Gerobatrachus*, which is informally called a "frogamander" and considered to be a possible missing link between the Paleozoic temnospondyls and modern lissamphibians (Figure 8.26). *Gerobatrachus* may be the common ancestor of frogs and salamanders. It was small (about ten centimeters long) and had a stubby tail and frog-like skull structure, but a more elongate body and without the comparatively strong back legs seen in modern frogs. By the Early Jurassic, true frogs and salamanders had finally diverged and came to occupy swamps and rivers across Pangaea.

TERRESTRIAL REPTILES

Although reptiles evolved during the Carboniferous, the Mesozoic is called the "age of reptiles" due to the diversity they achieved during this era. Dinosaurs are among the most famous of these reptiles but only constitute a subset of a much more diverse class of organisms that evolved in the Mesozoic, many of which are still with us today.

The earliest reptiles are generally called **protorothyrids** ("stem reptiles") because they are the common ancestors of all subsequent reptiles as well as birds and mammals (Figure 8.27). From these, several distinct lineages branched off and took very different evolutionary pathways.

Turtles and lizards evolved from Late Permian protorothyrids, and by the Triassic, they gave rise to the tuatara, which look like iguanas and are today only found in New Zealand. *Snakes* diverged from lizards sometime in the Late Jurassic or Early Cretaceous.

By the Triassic, some protorothyrids branched off and formed the **archosaurs** ("ruling reptiles"), a group that includes the extinct dinosaurs and flying reptiles (pterosaurs) as well as birds and crocodilians. Archosaurs came to dominate the land and air during the Mesozoic. They are classified as *diapsids*, which have openings (fenestrae) in their skull toward the backs of their heads (referred to as postorbital, meaning behind the eye sockets). These holes

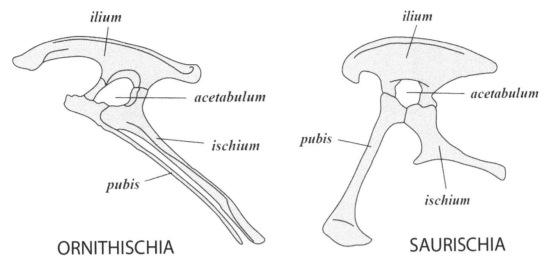

Figure 8.28: Left-side views of pelvic regions for Ornithischia (left) and Saurischia (right).

reduce the weight of the skull as well as provide better anchors for facial and jaw muscles. Archosaurs have one set of post-orbital fenestrae, while other reptiles typically have two on each side. Archosaurs typically have narrow skulls, pointed snouts, and socketed teeth. (Mesozoic birds did, in fact, have socketed teeth, but they lost these over time.) The most dominant archosaurs during the Mesozoic were the dinosaurs. Keep in mind that there were not any flying or aquatic dinosaurs; those that we know from the Mesozoic skies and seas were reptiles and birds, not true dinosaurs. Additionally, while most dinosaurs today are portrayed as very large (and many obviously were), many dinosaurs were no larger than a turkey. Most of the names of dinosaurs, including the term "dinosaur" itself, include the word "lizard" as a root. For instance, "dinosaur" means "terrible lizard." Dinosaurs were not lizards. They were reptiles and differed from true lizards in many ways.

DINOSAURS

The age of dinosaurs began with small, bipedal archosaurs in the Mid-Triassic. These began to show dinosaur-like traits, such as reduced fourth and fifth fingers/toes on their hands and feet, three or more vertebrae near the pelvis, and hip bones that connect in a way that forms a distinct opening for the top of the femur to attach and gives them a more erect stance. These common ancestors to all dinosaurs diverged into two general groups that can be differentiated based on the overall shape of the pelvis (Figure 8.28).

Ornithischian ("bird-hipped") dinosaurs have pelvises that resemble those of birds. The irony of this name is that these were not the dinosaurs that evolved into birds; it was the **saurischian** ("lizard-hipped") dinosaurs that birds diverged from in the Jurassic.

ORDER ORNITHISCHIA

This order includes a variety of dinosaurs that have hips shaped like those of birds (Figure 8.29). Ornithischian dinosaurs lack teeth in the fronts of their mouths and have a unique arrangement of rigid tendons in the backs of their jaws. They are subdivided into five distinct suborders.

1. Ankylosaurs: Meaning "fused lizards," ankylosaurs are most famous for their heavily armored backs and tails with a heavy club at the end. Like a turtle, many of their bones were fused together, making their

EXAMPLES OF ORDER ORNITHISCHIA

Ankylosaur

Triceratops

Pachycephalosaur
(*Dracorex hogwartsia*)

Iguanodon

Stegosaurus

Hadrosaur
(*Edmontosaurus*)

Figure 8.29abcdef: Examples of dinosaurs belonging to the order Ornithischia. Sizes not to scale.

bodies extremely strong and rigid. The armor on their backs was composed of osteoderms, plates with knobby protrusions similar to those of crocodiles and armadillos. They had two rows of spikes running down their bodies and horns protruding from the backs of their heads. These quadrupeds reached lengths of up to thirty-five feet but stood only six feet tall. To fend off predators, scientists think an ankylosaur would lay flat on its belly and swing its tail around. The rows of spikes along its side would help prevent predators from flipping it over to expose its more vulnerable underside. Ankylosaurs were herbivorous and lived during the late Cretaceous, about seventy-five to sixty-five million years ago.

2. Ceratopsians: These were a diverse group of compact, car-sized herbivorous dinosaurs that had large heads, beaks with grinding teeth toward the back of the jaw, and, with the exception of Early Cretaceous examples, had bony frills and horns on their heads. The most famous of these was the *Triceratops*, which, in addition to a large frill, had three horns pointing forward on its head and snout. The frill of ceratopsians may have been used for defense against other dinosaurs such as the *T. Rex*, but these frills are somewhat weak and may have rather served as heat regulators, competition for mates, or even for signaling other ceratopsians. Bone beds of ceratopsians in Asia and North America suggest they traveled in herds and possibly used stampeding to help fend off predators. Some species of ceratopsians that have been described may have been simply examples of different growth stages of the *Triceratops*. For example, the *Torosaurus* had been traditionally considered a separate species of ceratopsians, but more recent work has suggested that they were simply more mature versions of *Triceratops*. This theory is not universally accepted, as other scientists point out that there appears to be young versions of *Torosaurus* that look different from the standard *Triceratops*. The oldest ceratopsians appeared in the Early Cretaceous and diversified in the Late Cretaceous before going extinct at the end of the period.

3. Pachycephalosaurs: Somewhat related to ceratopsians due to their thick, bony heads, pachycephalosaurs fall under the same category of marginocephalia. Their name meaning "bone-head," pachycephalosaurs were bipedal herbivores with extremely thick skulls (up to ten inches thick in the front) fringed by bony protuberances and bodies about fifteen feet long. Their skulls are most commonly found, while the rest of their bones are rather rare in Cretaceous strata. It is assumed that they had a rather short, thick body and an upright stance. The purpose of the thick skulls may have been for head-butting during competition for mates (the males tend to have the thicker skulls), as well as for defense against carnivorous predators. Other genera of pachycephalosaurs have been described, but there is still uncertainty if there were, in fact, separate genera, or instead, examples of various growth stages of the same *Pachycephalosaurus* dinosaur. Regardless, other genera include *Dracorex hogwartsia* (Dragon King of Hogwarts), whose name was given by visitors to the Children's Museum of Indianapolis as an obvious shout-out to the *Harry Potter* books. *Dracorex* had the typical thick skull of pachycephalosaurs with a series of backward-facing spikes along the back edge of its head. Even more like a dragon, another genus, *Stygimoloch* (Demon from the River Styx), had two rows of progressively larger spikes running from its snout to the back of its head. *Stygimoloch* was only about ten feet long at best and on average was about as tall as a human. Pachycephalosaurs lived during the Late Cretaceous and were some of the last types of dinosaurs to go extinct.

4. Ornithopods: These were a diverse group of medium to large, mostly bipedal, herbivorous dinosaurs with well-developed forearms that could be used for quadrupedal locomotion as well. What sets them apart from the other groups is a rather large and complex sinus region of the skull, and in many cases, crests atop their heads. Many ornithopods are familiar dinosaurs and a few are described here. *Iguanodons* appeared in the Early Cretaceous, and their fossils are found throughout Europe and the United States. They were both bipedal and quadrupedal, depending on their preference; unlike the *T. Rex*, their front arms, like with most ornithopods, were strong enough to support their weight when needed. They had hands with four individual hoof-like fingers and a fifth arranged almost like a thumb with a large claw at the end. Probably not used for grasping, this "thumb" was more likely used for defense or for competition for a mate. Its tail was strong and rigid, with ossified (bony) tendons helping to keep the tail from dragging on the ground. *Hadrosaurs* are another group of ornithopods that include many of the "duck-billed" dinosaurs. Hadrosaurs are found in Upper Cretaceous strata across Europe, Asia, and North America. Like Iguanodons, hadrosaurs had stiff, raised tails to help with balance and maneuverability. Many hadrosaur nests have been discovered together, indicating they shared communal nesting grounds. One group of hadrosaurs, the *Lambeosaurinae*, had prominent crests with air pipes inside connected to their nasal passages; these crests were likely used to blow air through, creating a loud, deep, resonating noise used for communication. The other type of hadrosaur was called *Saurolophinae* and either lacked the crests on their heads or had crests that didn't have significant air chambers. Like Lambeosaurinae, they, too, had duck-billed mouths and often very large nasal passages. The *Parasaurolophus*, a North American hadrosaur belonging to the Lambeosaurinae, had a bony crest like its relatives and reached lengths of up to thirty feet. They likely dwelled within the fern-rich undergrowth of conifer forests. *Edmontosaurus*, on the other hand, belongs to the Saurolophinae and lived in regions that varied from the warm, Cretaceous seaway over North America all the way up to near the North Pole (which would have been warmer at that time than it is today). They had small, but strong, replaceable grinding teeth near the rears of their beaks and likely stripped tough leaves from branches when they ate. *Edmontosaurus* was one of the largest hadrosaurs, reaching lengths of up to forty feet.

5. Stegosauria: The best known genus of stegosauria, and the namesake, is the *Stegosaurus*, but this group also contains a plethora of other genera most closely related to the ankylosaurs. Like the other members of their group, stegosaurs were quadrupedal dinosaurs that reached lengths of over thirty feet. They are most famous for two anatomical traits: a set of spikes and/or flat, vertical plates that lined its back in two parallel rows and a tail ending with a set of large spikes, informally referred to as the *thagomizer*, a name given by Gary Larson, author of the *Far Side* comics. Evolving in the Middle Jurassic, stegosaurids first appeared in Asia, then evolved progressively larger and more ornate plates and spikes along their backs, culminating in genera such as the stegosaur (proper), which existed throughout North America in the Late Jurassic. Few known species made it into the Cretaceous. Although their tails were undoubtedly used for defense, the two rows of dermal plates lining their backs likely provided little protection from predators such as the *Allosaurus*, as they were thin and likely weak. Instead, they may have been used for

temperature regulation or simply as a way to help differentiate between members of a herd. Stegosaurids had some of the smallest brains of all dinosaurs, about the size of a walnut. Their heads were also small and narrow, often with a horn-covered beak.

ORDER SAURISCHIA

Saurischian dinosaurs are differentiated from the ornithischians discussed above in that their pelvic regions are more similar to those of lizards than to birds. However, only differentiating them by their hip shape isn't enough to appreciate the other major differences between these two orders. The saurischians are quite diverse, and as a result, they can be further divided into two major suborders: the **sauropods** and the **theropods**. Chronologically speaking, the saurischians showed up in the fossil record before the ornithischians, during the Triassic, and are thus more related to the earliest dinosaurs. They are described here for continuity's sake because it was from the saurschian theropods that birds evolved, and this transition will be discussed later in this chapter.

SAUROPODS

Sauropods (Figure 8.30) are the familiar, long-necked, long-tailed quadrupedal herbivores that include the largest and tallest creatures known to ever walk the planet. Sauropods evolved from some of the earliest dinosaur stock in the Triassic. Ancestors to sauropods were possibly bipedal, and by the Triassic, predecessors of sauropods, called prosauropods, had evolved and were semi-quadrupedal, showing the transition to the complete quadrupedalism of most sauropods. Prosauropods are some of the earliest true dinosaur fossils known so far. Their bones are found in Triassic strata from northern Europe as well as Madagascar and Antarctica. Being semi-quadrupedal, prosauropods had strong hind legs that likely enabled them to stand upright long enough to pluck leaves from high branches within the conifer forests they roamed in. Their teeth, made for grinding, additionally support the theory that these were herbivorous. They ranged in length from about ten to forty feet and had small heads with long necks and tails. Further evidence indicates they traveled in herds, likely for protection. Sauropods, in general, grew to full height within about ten years but had life spans that rival those of humans. This rapid growth rate may be an adaptation toward a migratory lifestyle, as the young would have to grow fast to be able to keep up with the herd.

The *Diplodocus* was one of the first true sauropods, appearing in the Late Triassic and lasting until the Cretaceous. They were

Figure 8.30: Examples of dinosaurs belonging to the order Saurischia, suborder Sauropoda. Sizes are not to scale.

BRACHIOSAUR

TITANOSAUR

CAMARASAURS

DIPLODOCUS

massive, reaching lengths of over one hundred feet and weights of ten to sixteen tons. Their long necks and tails made up much of their length. The tail of *Diplodocus*, compared to those of its prosauropod ancestors, had more vertebrae (up to eighty) and was used as a counterbalance to its long neck. Its tail was held outward and didn't drag; the flexibility of its tail enabled it to use it as a whip for defense, with some estimates suggesting the tail could reach a speed of eight hundred miles per hour, resulting in a sonic boom when employed.

Its head was small compared to its body and terminated in a set of pointed, rod-shaped teeth that jutted forward and were used to strip leaves from branches. Fossil skin impressions indicate a rough texture composed of small, pointy bumps. Its overall posture was horizontal, and its front legs, which contained a sharp claw, were a little shorter than its hind legs. Some scientists think they could have reared up onto their hind legs to reach tall vegetation, but this is debatable, as their hearts would have to be very strong to pump blood up that high to their heads. It is also theorized that they may have fed on aquatic plants by standing on land and reaching down below ground level into rivers and lakes using their long necks, or simply grazed land plants by swinging their long necks from side to side.

The *Apatosaurus* is a genus of dinosaurs traditionally referred to as brontosaurus; however, the name "*Brontosaurus*" has since been abandoned. Apatosaurs roamed North America during the Jurassic. They reached lengths of over seventy-five feet, weights of up to twenty-five tons and were more robust than the *Diplodocus*, with larger neck vertebrae and stronger leg bones. Its feet and overall anatomy were similar to those of *Diplodocus* and other sauropods; in addition, it likely had similar feeding habits and roamed similar habitats.

The *Camarasaurus* ("Chambered Lizard") was the most numerous of the sauropods and lived in North America during the Late Jurassic. They were close to fifty feet long and weighed up to eighteen tons, making them bulky, yet much shorter in length and with fewer vertebrae than apatosaurs and *Diplodocus*. Unlike the *Diplodocus*, the *Camarasaurus* had strong, thick teeth and gastroliths that helped it further digest tough plant matter. This means it could have coexisted with *Diplodocus* without having to compete for food. Camarasaur skulls are more blunt and tall than many other sauropods. They had nostrils and eyes placed high on their heads, a trait that at first implied they could move into bodies of water and use their nostrils as snorkels. Recent research has indicated this was unlikely, as the water pressure would make breathing very difficult. Camarasaurs were likely migratory and traveled in herds.

Brachiosaurs lived in Europe, North America, and North Africa during the Late Jurassic. They reached lengths of up to eighty-five feet and were rather robust, possibly weighing in at nearly fifty tons. Brachiosaurs had long necks (over thirty feet) relative to their bodies, more so than most sauropods, and had rather small heads. Like other sauropods, they likely couldn't raise their heads up high; their hearts weren't strong enough to pump blood vertically up their long necks. Instead, brachiosaurs and other sauropods likely kept their heads low to the ground.

They are different from many other sauropods in that their front legs were longer than their back legs, and their shoulders and the bases of their necks were higher up, possibly making it easier for them to reach higher branches than other sauropods could. Their skulls are small but wide, and they had spoon-shaped teeth that helped them crop or shear vegetative matter from conifers and other gymnosperms of the day. Anatomical evidence indicates brachiosaurs may have been warm-blooded.

Titanosaurs are a diverse and poorly understood clade of sauropods that included among them the heaviest creatures to walk on land, such as the *argen-tinosaur*, which was well over one hundred feet long

Figure 8.31: 1915 photo of a *Diplodocus* skull. Its teeth are designed to strip leaves off of branches.

and weighed close to one hundred tons. They were also some of the last sauropods to exist, hanging on through the Cretaceous until the very end. Although their fossil remains are scattered across every continent except for Antarctica, their bones tend to be found as isolated fragments, with no complete fossil unearthed so far. Titanosaurs are distinct from other sauropods in that they had an armored covering, likely in response to the plethora of large carnivorous dinosaurs. In most other respects, they fit in well with the typical sauropod body shape and lifestyle.

There is, or rather was, evidence of what may have been the largest dinosaur ever, *Amphicoelias fragillimus*. Only the top part of one vertebra was found in the 1870s, but it was so big that estimates indicate the entire vertebra (just a single bone) was nearly nine feet tall. Since this discovery, the bone has gone missing. As these dinosaurs are inferred to be similar to the *Diplodocus*, scientists are able to extrapolate the length of *fragillimus* as up to 200 feet long and a weight of 130 tons. These measurements are highly uncertain and based on the assumption that the *Amphicoelias* dinosaurs were analogous to the proportions of the *Diplodocus*. Other members of the genus *Amphicoelias* were smaller than this but still big by sauropod standards.

THEROPODS

Theropods ("beast foot") are the suborder of Saurischian dinosaurs that include many of the well-known, and largest, predatory carnivores (Figure 8.32). They had a long range, from the Late Triassic to the end of the Cretaceous. Theropods were bipedal and typically had greatly reduced forelimbs and hollow, delicate bones. Their hands and feet usually had three main clawed fingers and toes and two more reduced digits as well. Theropod teeth were generally sharp and curve inward, allowing them to tear flesh apart easily.

Theropods can be classified into one of four groups:

1. Herrerasauridae: The earliest of the three groups, typified by the *Herrerasaurus*, evolved in what is now South America during the mid- to Late Triassic. It is not yet known whether herrerasaurs were true theropods; they may have been transitional between theropods and earlier archosaurs. Herrerasaurs were generally small, quick, raptor-like carnivores with long, thin tails, small heads, and sharp, serrated teeth.

2. Ceratosauria: If the Herrerasaurs weren't the first true theropods, then ceratosauria were, as they, too, appear in Late Triassic rocks and were more similar to the typical bird-like theropods. Jurassic-age ceratosaurs, which lend their name to their clade, had lightweight, hollow bones and an S-shaped neck, both traits of modern birds. They were raptor-like theropods that reached lengths of up to twenty-five feet. Similar to these were the slightly smaller *dilophosaurs* (whose membership in the ceratosaur group is debated), which had prominent double crests on their heads.

3. Tetanurae: This group, the largest of the three, is subdivided into three clades: The Carnosauria ("meat-eating lizards") reached their peak during the Jurassic. These include some of the most famous meat-eating dinosaurs. Carnosaurs were agile, bipedal predators with short necks and long tails and are set apart from the other groups by a proportionally larger femur compared to their shin bone and large openings in their skulls, including those that housed the eyes. Members include the Jurassic *Allosaurus* ("different lizard"), which had larger arms (with more menacing claws) than the *T. Rex* (which came later), possibly allowing them to grab their prey before taking huge bites out of it (Figure 8.33). Allosaurs had a narrower skull with large jaws and sharp, serrated teeth. They were about thirty feet long, putting them at the top of the food chain, preying on ornithischians and even sauropods, which adapted by growing really big and eventually growing armor to protect themselves. Another carnosaur, the Cretaceous *Spinosaurus*, reached lengths of nearly fifty feet and weighed up

to nine tons, making them larger than the *T. Rex*. Their name is reminiscent of the long, bony spikes that formed a webbed sail down their backs. Living in warm African environments, their sails likely acted as a way to cool themselves or ward off competition. Recent computer models of the bones of *Spinosaurus* (most of the original bones were destroyed by allied bombing during WWII) indicate it may have been semi-aquatic, which is quite a novel idea. It's bones are dense and it's back legs appear to be designed for paddling. *Spinosaurus*, then, is not just the largest known of the carnivorous theropods, it is the only one known so far to spend part of its life in freshwater, likely hunting large fish in rivers.

The Coelurosauria ("hollow-tailed lizards") were even more bird-like than the carnosaurs. Many members of this group were feathered, and they generally had longer shin bones than femurs. They ranged in length from a couple feet to over forty feet. The tyrannosaurids ("Tyrant Lizards") were the largest and most fierce members of the coelurosaurs. They lived in North and South America and reached their peak at the end of the Cretaceous. Early tyrannosaurids were relatively small, but over time, they grew larger (up to forty feet long), their arms shrank and lost a digit (the *T. Rex* had two clawed fingers), and their skulls became more massive. Many tyrannosaurid fossils indicate they had feathers; these wouldn't have allowed flight but may have served as insulation or decoration.

4. Dromaeosauridae is a family that includes the "raptors," which were relatively small (from three to thirty feet long) and were bird-like in many ways, including feathers and long arms with a bony architecture that appears to be a prelude to the wing structure of birds, making them good candidates for the ancestor of modern birds. One of the most famous examples is the Late Cretaceous *Velociraptor* ("swift seizer"), which had long, curved claws on its hands and feet that it used

Figure 9.32: Examples belonging to the order Saurischia, suborder Theropoda. Sizes not to scale.

Figure 9.33 Neck and skull region of a *Tyrannosaurus rex*.

to slash at and grab onto its prey. Unlike the way they were depicted in *Jurassic Park*, they were only about six feet long, including their tail, and were likely covered in feathers. Another member of this family is the Early Cretaceous *Deinonychus* ("terrible claw"), which reached up to ten feet in length and nearly two hundred pounds (they were likely the inspiration for the velociraptors in *Jurassic Park*). The largest member of the Dromaeosaur family is the *Utahraptor*, which reached lengths of up to twenty-three feet and weighed up to fifteen hundred pounds. Like the other dromaeosaurs, they had a particularly huge claw on their hind feet, which in the case of the *Utahraptor*, reached over a foot long! It is theorized that dromaeosaurs would leap onto their prey, tearing deep gashes with their large hind claws, then waited for their prey to bleed to death before eating them with huge gulping bites.

Oviraptors were a genus of relatively small-to-moderate-sized theropods that were most dominant in the Cretaceous. Their name means "egg stealer," and was based on a particular oviraptor fossil that was found atop what was presumed to be a clutch of *protoceratops* eggs, and possibly died while trying to eat them. Later exploration of the eggs, which were very well-preserved, showed they were actually the eggs of the oviraptor. This discovery indicates the original fossil oviraptor was brooding over its own eggs instead of stealing. This discovery has two major implications; first, the oviraptor behaved like birds, protecting their eggs; second, the need to incubate may suggest that oviraptors were warm-blooded.

The shapes of their beaks and the presence, in some species, of a small spike on the ceilings of their mouths indicate they may have included eggs of other organisms in their regular diet. Other scientists believe they more likely used their strong beaks to crush the shells of mollusks. Oviraptors were, on average, about the size of a human, although they reached much larger sizes in some species. They had slender, elongate limbs that faced forward, and had wrists and fingers capable of grasping. Their beaks lacked teeth, but were muscular and capable of crushing shells (and possibly eggs). They had a small bony crest, like a reduced version of those of hadrosaurs. Overall, they appeared like large birds but with clawed, feathered arms instead of wings.

A recent and notable find was of a very large oviraptor named *Anzu wyliei*, but informally called the "chicken from hell." *Anzu* was about 10 feet tall and 11 feet long, and weighed in at about 600 pounds. Like other oviraptors, it had a crest on top of its head, and extremely sharp claws on its feet and forearms. So far, the limited fossil finds come from late Cretaceous beds, indicating this species likely went extinct not long after. The largest oviraptor known was the Late Cretaceous *Gigantoraptor*, which reached 26 feet in length and weighed up to two tons. Since its discovery in 2005, only fragmentary fossils of *Gigantoraptor* have been unearthed, so more fossil evidence is needed to elaborate on these giant oviraptors.

WARM-BLOODED OR COLD-BLOODED

It is still unclear whether dinosaurs were warm-blooded (endothermic) or cold-blooded (ectothermic), or if there were examples of both. Most anatomical evidence is inconclusive. However, many paleontologists would argue that the theropods were likely warm-blooded, while the sauropods may have been cold-blooded. Ornithischian dinosaurs also provide some evidence to suggest that they too, were warm-blooded. For instance, the duck-billed dinosaurs appear to have grown very quickly during their early years, a trait usually seen in endothermic organisms.

In addition, a particularly well-preserved ornithopod fossil contains evidence of a four-chambered heart, also implying it was endothermic.

FLYING REPTILES

There were no flying dinosaurs. Instead, the popularized flying "dinosaurs" are really flying reptiles, and differ from birds in many ways.

Flying reptiles compose the order *Pterosauria*, and are typically called pterosaurs (Figure 8.34). The word "pterodactyl" isn't a real scientific designation, and won't be used in this book otherwise. Pterosaurs first show up in the fossil record in rocks from the Late Triassic (making them the first flying vertebrates), and were around until the end of the Cretaceous. The fossil record is sketchy in regards to the origins of pterosaurs. The age of their fossils indicate they evolved *alongside* the first true dinosaurs, so their common ancestor is likely an earlier early archosaur, although some paleontologists place their roots among the earliest dinosaurs.

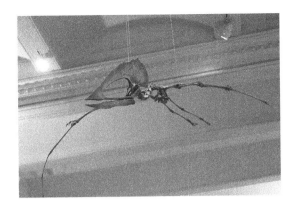

Figure 8.34: A fossil of a pterosaur (*Tupuxuara leonardi*), at the American Museum of Natural History, New York City.

Pterosaurs had relatively large limbs and heads compared to their short torsos. Although they are often depicted as frail, light-weight creatures, they were quite muscular, and their wings (forelimbs) were especially robust. The wing membrane was composed of skin, muscle, blood vessels, and stiff, semi-rigid fibers for extra support. The wings connected to both the front and hind limbs. The main bone in their wing was an extended fourth finger. Larger examples likely glided on rising thermal air plumes like hawks, while the smaller pterosaurs probably flapped their wings much more to maintain flight. Pterosaurs had hollow, light-weight bones filled with air sacs connected to their respiratory systems, like those of birds. To make up for the extra weight of their muscle mass, their bones were thin-walled and designed to be as light as possible. Their skulls, for instance, had numerous fenestrae. Pterosaurs ranged greatly in size, with wingspans between several inches and nearly 40 feet. Most pterosaurs were bipedal while on the ground, although some of the larger examples may have been quadrupedal, walking on its winged forelimbs. Besides their size, what varied the most was the shapes of their heads, which appear to have been adapted towards a variety of food sources.

Pterosaurs are divided into main groups. The early basal forms, previously referred to as *rhamphorhynchoids*, include specimens such as *Dimorphodon*, a relatively small pterosaur (wingspan of about 5 feet) with large claws on its hind legs and wings that allowed it to climb around on trees and rocks quite well. Its body was rather robust, and it may not have been a very good flier, likely relying on short bursts of flight to escape predators. *Dimorphodon* had sharp, pointy teeth in the front of its jaw and more blade-like teeth towards the back, indicating it probably had a diet of insects and/or small vertebrates. Another basal pterosaur, *Rhamphorhynchus*, was a fish-eater, as some of their well-preserved fossils contain fish bones in their abdomen (Figure 8.35). This fur-covered flying reptile was

Figure 8.35: Cast of a fossil of *Rhamphorhynchus* from the Oxford University Museum of Natural History. The head was missing from the original fossil and has been added for the cast.

Figure 8.36: Model of *Quetzalcoatlus* from the Staatliches Museum für Naturkunde, Karlsruhe, Germany.

even smaller than *Dimorphodon*, with a wingspan of about 3 feet. Its jaws were narrow, long, and lined with sharp, pointy teeth that jutted outward, allowing it to possibly glide just above the water and scoop its mouth down to collect fish, similar to the hunting habits of modern birds such as the pelican. As the Mesozoic progressed, pterosaurs diversified.

A newer group of pterosaurs, called *Pterodactyloids*, dominated the Late Mesozoic. One of the members, Pteranodon, is a well-known pterosaur with wingspans that reached over 20 feet. Pteranodons lacked teeth altogether, and had a large bony crest towards the back of its head. They were marine feeders, scooping up fish as they glided above the water. One of the largest pterosaurs was *Quetzalcoatlus* (named after an Aztec god), which had a wingspan of up to 40 feet (Figure 8.36). While on the ground, *Quetzalcoatlus* appeared similar in size and posture to a giraffe, walking on all four limbs, and possibly took flight by either flapping its extremely long and muscular wings, or even by jumping off of hills or cliffs like hang-gliders do. Some scientists think that *Quetzalcoatlus* may not have flown at all, having bodies too heavy to fly efficiently, especially if it was cold-blooded, which is debatable. If it flew, it may have skimmed bodies of water for fish, but if not, it likely walked around and plucked small mammals or dinosaurs from the ground for food.

OTHER LAND REPTILES

As mentioned earlier, crocodilians had evolved in the Early Mesozoic from the stem reptiles, and by the Jurassic, came to dominate freshwater swamps and rivers. They appeared then very similar to their modern (extant) descendants, except that in the Cretaceous, they had reached lengths of nearly 50 feet. Turtles have also remained more or less the same since the Triassic, although the Mesozoic examples reached larger sizes than those today, up to ten feet long (Figure 8.37). Lizards, and their evolutionary off-shoots, snakes, also existed in the Mesozoic, although snakes didn't appear until the Late Jurassic or Early Cretaceous.

TRANSITION TO BIRDS

Although the link between theropod dinosaurs and birds is a recent theory, the fossils of some of the earliest birds had been discovered as far back as the 1860s, and sat in storage for well over a century before a new generation of paleontologists studied them in the context of bird evolution. A more recent slough of fossil finds from China has added even more essential details to the fossil record of early birds. It appears that theropod dinosaurs are the direct ancestors of birds, and many anatomical and physiological parallels can be drawn between the two. Bird-like features sprang up in theropod dinosaurs throughout the mid-to late-Mesozoic. Hollow bones with air sacs connected to the respiratory system, typical for birds, composed the 213-million-year-old fossil of *Tawa hallae*. The appearance of feathers, such as those on oviraptors, tyrannosaurids and dromaeosaurids, was one of the next

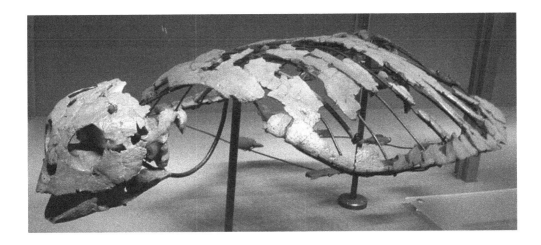

Figure 8.37: Fossil of *Allopleuron*, a type of turtle that lived from the Cretaceous to the Eocene.

signs of bird evolution. Oviraptors, like birds, practiced egg brooding. Meanwhile, the feet of many theropods had evolved and lost the fourth and fifth digit, changing to a bird-like form. It appears that many distinct lineages of birds had evolved separately and were truncated by extinction, especially at the end of the Cretaceous.

There is debate about the earliest birds, with some paleontologists placing them in the Late Triassic, such as *Protoavis*, which existed 80 million years before the famous *Archaeopteryx*. This contentious theory indicates the ancestors of birds were not the coelurosaurian dinosaurs, as previously thought. Unfortunately, many assumptions must be made about *Protoavis*, whose bones were found scattered and may not have been pieced back together correctly. Although it shows many bird-like features, the fossil lacks feather imprints and it may simply be the product of convergent evolution between an early dinosaur and more recent birds. Generally, scientists place the direct theropod ancestors of birds under the Maniraptoran ("seizing hands") clade, which includes, among others, dromaeosaurs and oviraptors.

The earliest fossil specimen regarded by most paleontologists as a true "bird" is the 150-million-year old *Archaeopteryx* ("ancient wing"), whose first fossils were found in Upper Jurassic strata in Germany by paleontologists over a century and half ago (Figure 8.38). Between 7 and 10 skeletons of *Archaeopteryx* have been found so far, and they are mostly in excellent condition and a large amount of information has already been gleaned from them.

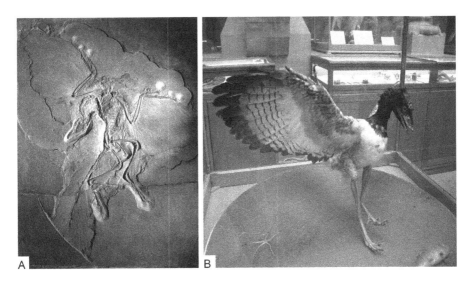

Figure 8.38: A) *Archaeopteryx lithographica*, a famous specimen displayed at the Museum für Naturkunde in Berlin, Germany; B) Model at the Oxford Museum of Natural History.

Figure 8.39: Fossil of *Confuciusornis sanctus* at the Hong Kong Science Museum.

Archaeopteryx was about the size of a seagull and had a small skull with long, sharp teeth, and a covering of long, slender feathers over most of its body, including its long, bony tail. Its "wings" still maintained three sharp-clawed fingers and were possibly used for grasping and climbing tree trunks. Its feet also had a hyperextensible second toe with a large claw, similar to those of other raptor-like dinosaurs.

Although its wings contained remarkably long feathers, it is still under debate whether they were capable of true flying. Most flying birds today have a large, protruding sternum upon which the flight muscles could attach. The sternum of *Archaeopteryx* is flatter and less pronounced, indicating that if it was capable of sustained flight, it was likely through gliding, and possibly short bursts of flapping. It likely also used its wings for brooding its eggs.

There are two hypotheses about the development of flight in early birds. One is the "from-the-trees-down" hypothesis, where birds first began to use their wings to glide down from high branches they had climbed to. The other hypothesis, the "from-the-ground-up," indicates flight initiated as short leaps into the air to catch flying insects or to avoid predators, and also as a way to add a burst of speed while climbing steep inclines.

A contemporary of *Archeopteryx* was the *Confuciusornis*, whose ubiquitous crow-sized fossils show up in Early Cretaceous strata from China (Figure 8.39). Some groups of fossils are so densely packed, they suggest that whole flocks of *Confuciusornis* died at the same time, possibly from nearby volcanic eruptions. Unlike the *Archaeopteryx*, *Confuciusornis* lacked teeth, and is the oldest bird known so far to have a true "beak." In addition, it lacked a longbony tail, and instead had a short, fused set of vertebrae like those of modern birds. Two long tail feathers, however, appear to have been attached, as shown in many fossil examples. Like the *Archaeopteryx*, *Confuciusornis* likely wasn't a good flier. Its bone structure in its wings suggested it couldn't raise them above horizontal, precluding flapping flight, which requires more flexibility. Furthermore, its feathers may have been too weak for efficient flight. Some evidence indicates *Confuciusornis* could better use its clawed wings to climb the branches of trees. Some fossils of *Confuciusornis* were associated with fish bones, suggesting they hunted along bodies of water.

The Mid- to Late Cretaceous witnessed a rapid diversification of birds. *Enantiornithes*, found in strata ranging from 135 to 65 million years ago, were the most diverse lineage of birds in the Mesozoic. It is still debatable whether these were closely related to *Archaeopteryx*, or belonged to a completely separate lineage of birds. *Enantiornithes* were good fliers, and lived mostly arboreal lives, with some swimming/wading species living near water. Like many early birds, *Enantiornithes* had teeth. Their wingspans ranged in size from several inches to several feet.

Another lineage of toothed birds called *hesperornithiformes* ("western bird") evolved in the Cretaceous and became aquatic birds, incapable of flight, but quite agile in the water, only leaving it to lay eggs and brood (Figure 8.40). They were excellent swimmers with thick, furry insulating feathers, and likely dove into the water to catch fish. They reached lengths of about 4 feet. Another group of Cretaceous birds to adapt to an aquatic lifestyle were the pigeon-sized *ichthyorniformes*, which lived and looked much like modern-day seagulls and other water skimmers. They had backward-curving teeth towards the backs of their beaks that were likely used to pluck fish from the water. Unlike the *hesperornithiformes*, *ichthyorniformes* lived mostly on land and skimmed the water while hunting.

By the end of the Cretaceous, many shorebirds resembled the geese, ducks, and albatrosses we see today, while many others, like *Archaeopteryx*, stayed mostly on land and maintained a more theropod-like posture. It is unlikely that modern birds (subclass Neornithes) arose from any of those discussed above, with the possible exception of *Archaeopteryx*; the others were separate and diverging lineages of birds that were successful in the Cretaceous but weren't likely the *direct* ancestors to modern birds.

Figure 8.40: Classic painting by Heinrich Harder (1858–1935) of hyesperornithiformes.

The *Archaeopteryx* is the earliest known species belonging to the class Aves, which includes the subclass containing modern birds, Neornithes. Neornithes appear in the fossil record within Cretaceous strata (the earliest is the 85-million-year-old chicken-like *Austinornis*), and it thus appears that the direct ancestors of modern birds, both flightless and avian, had already evolved before the mass extinction event that capped the Cretaceous. The exact evolutionary links between the variety of Cretaceous birds and Neornithes (modern birds) is still unclear in the fossil record. Keep in mind that the bones of birds are especially thin and frail, and don't fossilize well.

TRANSITION TO MAMMALS

While one group of reptiles slowly gained avian characteristics and evolved into birds, another group was heading in a different direction, and eventually evolved into mammals. This transition into mammals began in the Late Paleozoic, when *pelycosaurs*, fin-backed reptiles, evolved from the earlier stem reptiles. A group of carnivorous pelycosaurs gave rise to a diverse and populous lineage of reptiles called *therapsids* about 275 million years ago, which had, by the end of the Permian, evolved many mammalian traits. These included: a fusing of many bones in the skull, giving them larger and less numerous skull bones compared to other reptiles; a more robust jaw with different types of teeth, such as those for biting, tearing, and grinding food, and; a more erect quadrupedal posture with longer middle toes than the first and fifth.

Therapsids diverged into several distinct lineages, some of them surviving the Permian Extinction and lasting until the end of the Mesozoic. Each of these groups became progressively more mammal-like, but only one group, the *cynodonts*, can be traced to today's mammals.

A notable early type of therapsid (not part of the direct mammalian lineage) were the herbivorous *dicynodonts* (Figure 8.41), which first appeared in the Permian and continued into the Mesozoic, surviving the worst mass extinction ever. Dicynodonts ("dog tooth") ranged from rodent- to cow-sized, and had a unique beaked snout that

Figure 8.41: Illustration of a typical dicynodont.

Figure 8.42: Illustration of a Gorgonopsid, a theriodont therapsid.

Figure 8.43: *Charassognathus*, a basal cynodont.

Figure 8.44: Illustration of the Late Cretaceous triconodont, *Repenomamus giganticus*.

included, in many species, a pair of short tusks. Larger dicynodonts had upright hind legs, but their front legs sprawled, bending at the elbows.

Another group of therapsids, the *theriodonts* ("beast-tooth"), gave rise to cynodonts, the direct ancestors of mammals, in the Late Permian. Theriodonts were arguably the most mammal-like of all the therapsids, especially in the way their jaws were constructed, giving them a wide gape. One type of theriodont, called *Gorgonopsia*, was built like a sabre-toothed cat, including a pair of large canine teeth (Figure 8.42). Some grew as big as bears, and most were agile and fast runners, likely hunting dicynodonts for prey.

By the Late Triassic, true mammals had evolved from cynodonts. The most basal cynodont currently known is the *Charassognathus* (Figure 8.43), a rodent-sized creature with a long snout and a set of large canine teeth. The earliest true mammals in the fossil record are differentiated from their therapsid ancestors mostly by the architecture of bones in their skulls. Mammals, unlike reptiles, have three, instead of one, bones in the middle ear. The extra two ear bones in mammals are derived from the bones in reptiles that connect the jaw to the skull. This evolution of the incus and malleus of the inner ear from what used to be jaw bones was slow and progressive, the stages represented well in the fossil record, as well as during the embryonic stages of all modern mammals, including humans. Cynodonts also began to show double-rooted, differentiated teeth that were replaced only once. Reptiles, on the other hand, constantly replace their teeth. Additionally, cynodonts evolved to have fused bones forming the top of the mouth (the palate), which separates the mouth from the nasal cavity allowing for chewing and breathing at the same time, a multitask well-suited for warm-blooded organisms. A rare fossil showing the skin impression of a cynodont indicates it had skin more like those of mammals than reptiles. Some cynodont fossils are found in complex, interconnected burrow systems, indicating at least some of them lived subterranean lives.

Throughout the Mesozoic, mammals remained primarily small (mouse to rat sized), although some got so big as to be able to eat baby dinosaurs, such as the Early Cretaceous mammal *Repenomamus giganticus*, which was over three feet long (Figure 8.44). The first mammal known to have fur was the semi-aquatic *Castorocauda*, from the Mid-Jurassic, and its fur was well-developed, including an insulating layer of underfur (Figure 8.45). This implies fur likely developed even earlier, but its first arrival hasn't shown up in the fossil record yet. *Castorocauda* also indicates that various mammalian traits showed up at different times, indicating that the transition from reptiles to mammals was a form of mosaic evolution.

By the Jurassic, several lineages of mammals had diverged from the cynodonts, including two lineages that connect with modern mammals. The *Eupantotheres* evolved in the Jurassic and branched off into Marsupial and Placental mammals during the Cretaceous. The *triconodonts* (named for the three cusps on their molars) also evolved in the Jurassic and gave rise to the monotremes (egg-laying mammals) in the Cretaceous as well (Figure 8.46). The only surviving monotremes are the platypus and spiny anteaters (echidnas). Placentals, which compose a clade referred to as *Eutheria*, became the dominant mammal during the later part of the Cenozoic, crowding out and leaving behind a much smaller group of marsupials that survive today, mostly in Australia. Another significant mammalian order, the *Multituberculates*, evolved in the Late Jurassic or Early Cretaceous, but went extinct in the Oligocene due to competition from rodents (Figure 8.47). Mesozoic mammals had relatively large brains considering their body size, and the portions of their brain devoted to the senses of smelling and hearing were well-developed, which may indicate a primarily nocturnal lifestyle. This likely reduced competition with dinosaurs for food, as well as reduced the chances of being eaten, as most dinosaurs were likely diurnal (awake during the day).

The earliest known mammal in the placental lineage was the 125-million-year-old *Eomaia* ("ancient mother"), a shrew-like, furry creature that, like many rodents of today, spent its time navigating the underbrush of forested areas. It was not a placental itself, but appears to have been part of their ancestry. The oldest marsupial known in the fossil record is the *Sinodelphys*, which was a contemporary of *Eomaia* (Figure 8.48). It was smaller, about the size of a chipmunk, and hunted insects in low-lying undergrowth and

Figure 8.45: Illustration of the semi-aquatic *Castorocauda*, one the earliest mammals known to have fur.

Figure 8.46: Fossil of triconodont (*Gobiconodon ostromi*) at the American Museum of Natural History, New York City.

Figure 8.47: Skull of a Cretaceous multituberculate, *Meniscoessus*, at the Rocky Mountain Dinosaur Resource Center, Woodland Park, Colorado.

branches. Although monotremes appear the most similar to reptiles in the context of birthing habits, their earliest fossils (that have been found so far) appear in strata over 25 million years *younger* than those containing the earliest placental and marsupial mammals. Regardless, monotremes, to this day, have a more primitive sprawling posture than other mammals. Fossils of these early monotremes, such as *Steropodon*, are fragmentary and in poor condition (Figure 8.49). It is theorized, however, that these cat-sized monotremes were similar in shape to the modern platypus and were some of the largest true mammals known in the Mesozoic.

Figure 8.48: A) Fossil of *Eomaia*; B) Illustration of *Eomaia*.

Figure 8.49: *Steropodon*, an Early Cretaceous monotreme.

Figure 8.50: A fossil cycad (*Zamites feneonis*) branch from the Upper Jurassic.

MESOZOIC PLANTS

By the start of the Mesozoic, the lycopsids and sphenopsids, which were prevalent during the Paleozoic, had been reduced to small, ground-covering plants, and the end of the Triassic witnessed the extinction of the seed ferns. The gymnosperms, however, continued to thrive into the Mesozoic, even evolving into new forms, such as the *cycads*, during the Triassic. Cycads appear superficially similar to ferns, but produce a cone of seeds, like most gymnosperms do (Figure 8.50).

Another type of plant, angiosperms, also evolved, perhaps from the gymnosperms as far back as the Triassic, as some research suggests. Angiosperms are flowering plants, with seeds that are protected by some type of fleshy covering, such as fruit. They also have diagnostic patterns of veining in their leaves; smaller veins branch off from a central axial vein, but instead of ending at the edge of the leaf, the veins loop back around and reconnect with the central vein (Figure 8.51).

The origins and early history of angiosperms is still unclear, but fossil record shows that by the mid-Cretaceous, they had diversified into many forms. The lack of fossil evidence pertaining to the origins of flowering plants was even declared an "abominable mystery" by Charles Darwin.

Angiosperms were, and are, so successful in adapting to their environment that they constitute 90% of all plant life today (up to 300,000 species), and grow in almost every type of terrestrial ecosystem imaginable. Angiosperms co-evolved with pollinators like bees and moths, which play an integral role in fertilization (Figure 8.52). Pollinators, such as insects, carry pollen from male flowers and deposit them on female flowers; here, the pollen sends tubes down into the stigma of the female flower, where the egg is fertilized and eventually grows into a seed. Often, the seeds remain within the swollen ovary of the plant, known as a *fruit*, allowing the seeds to be dispersed by organisms that eat it.

Evidence of the earliest angiosperms lies in the presence of pollen found in Triassic and Jurassic rocks. Drilling cores from Switzerland recently unearthed pollen grains embedded in 240-million-year old sediments, which may be the earliest fossil evidence of pollen, and thus, flowers. A long-accepted theory is that the *earliest* angiosperms grew in arid environments where they would not fossilize well, hence their paucity in the fossil record. Throughout the Early Cretaceous, angiosperms diversified, but were still limited to small populations along streams, and competed with ferns and cycads for space.

Another contentious aspect of angiosperm evolution is whether they were derived from trees and other woody shrubbery or rather from smaller herbaceous plants. The *Woody Magnoliid Hypothesis* states that the earliest angiosperms were derived from trees similar to today's magnolias and laurels. This theory is also supported by molecular research, indicating an ancestry of understory trees and bushes. The *Paleoherb Hypothesis*, on the other hand, states the earliest angiosperms were tropical herbaceous plants. In either case, it appears that flowers and pollen were not the first trait to define angiosperms; it was rather the vein structure in leaves, implying a mosaic-like series of evolutionary steps leading to angiosperms. The fossil record, thus far, indicates the evolution of angiosperms, although it occurred in steps, took place over a short period of time during the Early Cretaceous.

The oldest flower fossil (more than just the pollen), belongs to *Archaefructus liaoningensis* (Figures 8.53 and 8.54). It dates to about 142–125 million years ago (dating results vary), during a period believed by many paleobotanists to be marked by a major radiation and diversification of angiosperms. The fossils appear to have formed when plant matter fell into a pond and was buried by soft sediments. They are so well-preserved that the roots, leaves, flowers, and seeds are still intact and in their appropriate arrangement. The leaves were thin and lacy, and the stems were elongate and lined with both male and female flowers.

A fossil angiosperm, *Potomacapnos apeleutheron*, from 125–115 million years ago was recently rediscovered in the collections at the Smithsonian Natural History Museum, and is, so far, the oldest known angiosperm found in North America. The fossil was first believed to be part of a fern, but as the upper layer of sediments were chipped away, broad leaves were uncovered. The looped-vein pattern in the leaves are indicative of angiosperms and the shapes of the leaves are similar to those of modern poppies.

A 100-million-year-old piece of amber from Burma contains a very well-preserved bunch of tiny flowers belonging to *Micropetasos burmensis*, frozen in time during the act of fertilization (Figure 8.55). The pollen had been collected on the stigma of the female flower, and was in the process of sending tiny pollen tubes into it for subsequent germination. This specimen is one of the most intriguing and well-preserved fossil flowers known so far, and is the oldest direct fossil evidence of angiosperm reproduction.

Figure 8.51: Leaf of a ficus tree, showing a distinctive closed-loop pattern of veining typical of angiosperms.

Figure 8.52: A bee is covered in pollen while feeding on nectar from this flower. It will spread this pollen to the next flowers it visits.

Figure 8.53: A fossil of *Archaefructus liaoningensis*, the earliest known fossil flower (so far).

Figure 8.54: After an apple blossom (top) is fertilized, the ovary of the tree swells around the fertilized seeds to form an apple (bottom). The apple is an adaptation to attract animals to eat them and scatter the seeds.

By the end of the Mesozoic, a wide variety of angiosperms were on their way to terrestrial domination. The gymnosperms, which had added ginkgo and cycads to their group, also continued to thrive alongside ferns and mosses. As the climate warmed and became more tropical during the Mesozoic, lush forests spread across much of the continental land mass. The tallest trees were likely composed of a variety of conifers, while the understory was composed of tree ferns, seed ferns, and cycads. The undergrowth was dominated by ginkgoes, ferns and lower-growing cycads, as well as the early angiosperms beginning to claim their rightful place among the flora.

THE CRETACEOUS-TERTIARY (K-T) MASS EXTINCTION

Around 65–66 million years ago, a spectacular sight to behold lit up the skies of the Western Hemisphere and sent a wave of destruction across the planet (Figure 8.56). A six-mile-wide asteroid collided with the Earth near what is now the Yucatan Peninsula with a speed of over 12 miles a second, creating an explosion equivalent to millions of nuclear bombs. At the moment of collision, both the asteroid and about 90,000 cubic kilometers of the surrounding crust disintegrated and were blown upwards in the form of a growing, billowing cloud of incandescent dust and vaporized rock and metal whose top sent material clear into space. The sheer intensity of the light created by this impact would have heated the air to boiling temperatures and ignited fires across thousands of miles within seconds. Giant concentric ripples rolled through the crust away from the impact, a thousand times stronger than any possible fault-based earthquake, causing landslides across the landscape, forming sinkholes, and tossing rocks, boulders, and animals into the air. A shock wave would have spread through the air and knocked down most of the trees across eastern North and South America. Next came a giant tsunami, over 1,000 feet tall, that raced onto land and killed every living creature in its path. Droplets of glass called spherules and tektites, rained down from the skies near the impact site, at this point a glowing crater over 100 miles across.

Figure 8.55 Amazingly preserved fossil of *Micropetasos burmensis* in amber, trapped while being fertilized by pollen. (credit: George Poinar, "Micropetasos burmensis in amber being fertilized," http://oregonstate.edu/ua/ncs/archives/2014/jan/amber-fossil-reveals-ancient-reproduction-flowering-plants. Copyright © 2013 by George Poinar. Reprinted with permission.)

Within a radius of several thousand miles, nearly 90% of life was likely killed.

A dark cloud of dust, smoke, and ash circled the entire Earth within a few days and blocked out the sun for months, creating a severe global winter. The contaminated atmosphere became rich in sulfuric acid (H_2SO_4) and nitric acid (HNO_3), resulting in acid rain that immediately began to destroy vegetation. Furthermore, with such little sunlight reaching the surface, plants began to die en masse across the planet. The severe reduction of vegetation (about 60% of plants went extinct) affected the herbivores first, especially the larger ones, and they slowly began to starve. With fewer herbivores, carnivores began to die in large amounts. After the dust settled, nearly half of the Earth's species had gone extinct, many of them in the oceans. Nearly all the larger-sized vertebrates went extinct, including the non-avian dinosaurs, mosasaurs, plesiosaurs, and pterosaurs. In the oceans, ammonites, belemnites, rudist bivalves, and many types of plankton and reef-dwelling invertebrates (especially in tropical regions) went extinct. On the other hand, many groups of animals, such as insects and arachnids, mammals, birds, corals, mollusks, and angiosperms made it through relatively unscathed, or at least, rebounded quickly. After the dust settled, huge amounts of CO_2 in the atmosphere would have resulted in a period of warming once the global winter ended.

The asteroid impact theory, described above, first took hold in the late 1970s when geologist Walter Alvarez and his father, Nobel Prize–winning physicist Luis Alvarez, described a clay in Italy marking the end of the Mesozoic. Known as the Gubbio clay, this 2-3 cm–thick layer is enriched in iridium, an element rare on Earth but common in asteroids and meteorites (Figure 8.57). The level of iridium within this clay layer is 30 times richer than average. This same iridium-rich horizon has been found in over 100 sites (and counting) across the globe. This clay layer also contains microscopic droplets of glass, called spherules, that would could have only been produced by a large asteroid impact. The size of these droplets are largest in and around the Caribbean, where in 1991, geologists found evidence of a large crater dating to the K-T event. Drill cores and magnetic anomaly measurements provide evidence of a 110- mile-wide crater (known as the Chicxulub Crater) off the coast of Mexico's Yucatan Peninsula (Figure 8.58). Rocks from this crater as well as the K-T layer contain droplets of glass from melted rocks. In addition, they contain crystals of shocked quartz that have been bent and kinked from the shock-waves that penetrated into the crust during the impact. Similar crystals are formed beneath large nuclear blasts. In many locations, especially near the Caribbean, the K-T layer is found right beneath a layer rich in carbon and soot, indicating a period of intense

Figure 8.56: Illustration of the asteroid impact at the end of the Cretaceous.

Figure 8.57: The iridium spike is present within the light clay layer in this section of strata from the Cretaceous / Tertiary Boundary in Wyoming.

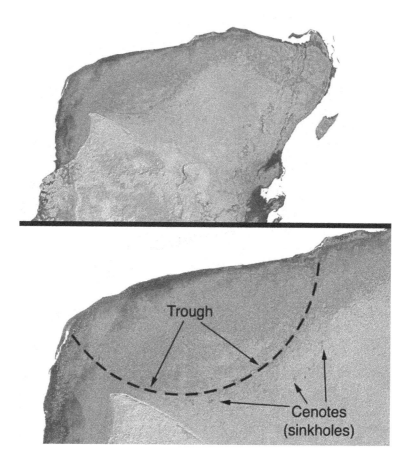

Figure 8.58: NASA radar-based image showing topography of the Yucatan Peninsula and landforms associated with the deeply buried crater, such as a ring of sinkholes surrounding the impact site.

forest fires. Furthermore, evidence of a large tsunami, in the form of large boulders on what used to be the ocean floor, can be found in places such as the Brazos River basin in Texas.

There are several lines of evidence that support the theory that an asteroid impact occurred at the K-T boundary:

1. The consistent presence of iridium, shocked quartz, and glass in the K-T clay layer.

2. The age of the Chicxulub Crater makes it the best candidate for the impact site.

3. A scattering of random boulders across the Caribbean 65 million years ago, indicating a large tsunami tumbled them across the ocean floor and onto land.

4. Asteroids this size are ominously numerous in our solar system.

Although it is quite evident an asteroid hit the Earth a little over 65 million years ago, one cannot rule out other theories about what caused the extinction event. During this period of time, prolonged, large-scale eruptions of basalt were taking place in India from about 68–60 million years ago (known as the **Deccan Traps**). This would have likely caused similar climatic swings to those caused by the impact, but over a longer span of time. The climate record from the K-T boundary is not resolved in good detail, but it appears the impact was preceded by a period of climatic instability. Some researchers think many of the dinosaurs were already on their way out prior to the impact. Whatever the cause or causes of the extinction were, it is very likely the asteroid impact was a major contributor to the event.

Figure 9.1: Open grassland at Shawangunk Grasslands National Wildlife Refuge in New York State. Ecosystems like this replaced dense forest across much of the mid-latitudes during the Oligocene.

CENOZOIC LIFE

CENOZOIC GEOLOGIC HISTORY

By the start of the Cenozoic, Pangaea had, for all intents and purposes, broken up, although the arrangement of continents was a little different from that of today (Figure 9.2). The Atlantic Ocean had already formed by the rifting apart of the Americas from Africa and Eurasia but was not as wide as it is now. North America was divided by the Cretaceous Interior Seaway and was not yet connected to South America. Another large interior seaway divided Europe from the rest of the Eurasian Plate. India was rapidly drifting northward and eventually crashed into Asia by the Eocene. The Middle East was still connected to Africa, and a shallow sea separated Africa from Europe to the north. Australia detached from Antarctica and drifted North toward lower latitudes and milder climates. Sea levels were extremely high, submerging large areas of the continents beneath shallow epeiric seas.

Two major belts of mountain-building and volcanism formed during the Cenozoic. One was the development of the Pacific Ring of Fire, more formally known as the circum-Pacific orogenic belt. This is a series of subduction-zone plate boundaries that formed along the western edges of North and South America, the North Pacific, specifically the Aleutian Island arc, the eastern edges of Asia, and the land masses in the South Pacific, such as Australia and New Zealand. Although some of these convergent boundaries formed during the Mesozoic, the Cenozoic witnessed continuing activity along these zones.

The other major mountain-building event was the formation of the Alpine-Himalayan orogenic belt. As Pangaea broke apart, the Tethys Sea, which occupied a large indentation along Pangaea's eastern coastline, began to close up as the African, Arabian, and Indian Plates drifted northward. The zone of collision between these plates and Eurasia is delineated by the still-active Alpine-Himalayan orogenic belt, named after the two large mountain ranges that make up much of the boundary. Deformation from this boundary began in the Mesozoic and has continued to this day. The only modern vestige of the Tethys Sea is the Mediterranean, which is a basin produced by the complicated tectonic upheavals during the formation of the Alps, Apennines and Pyrenees mountain ranges of Europe and the Atlas Mountains of northwestern Africa during the Alpine orogeny.

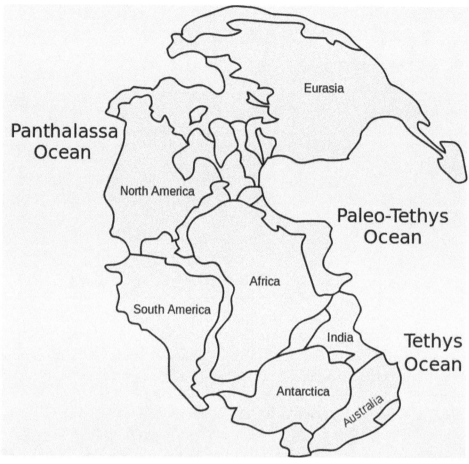

Figure 9.2: Pangaea in the Late Mesozoic.

To the east, the Indian Plate was still drifting north, and the subduction of oceanic crust beneath Asia resulted in a volcanic arc in what is now Tibet. This volcanism waned when the intervening sea closed up and the continent of India finally collided with Asia during the Himalayan orogeny during the Eocene. Since both India and Asia are composed of continental crust, niether subducted when they collided. Instead, a series of large-scale thrust faults created a wide and extra-thick portion of crust constituting the Tibetan Plateau and Himalayan mountain range.

Along western North America (called the Cordillera), a series of orogenic events throughout the Mesozoic added landmass and created wide belts of mountains and volcanoes. This was caused by the subduction of the Farallon Plate beneath North America; today, the only remnants of the Farallon Plate are the Juan de Fuca and Cocos plates (Figure 9.3). A divergent boundary between the Pacific and Farallon plates migrated east across the Pacific, causing rocks of the subducting Farallon Plate to be progressively younger in age and therefore more buoyant and harder to subduct. As the oceanic crust continued to shove beneath North America, the angle of subduction became progressively shallower. This caused an eastward migration of mountain-building that deformed much of western North America, culminating in the Laramide orogeny, which began in the Late Jurassic and continued into the Cenozoic. The Laramide orogeny (Figure 9.4) caused a series of fault-bounded blocks of crust to uplift as far east as the Rocky Mountains, which are, in part, the product of this orogeny. The period of mountain-building in the West waned by the middle Eocene.

A plume of heat beneath the Cordillera continued to cause volcanism west of the Rockies throughout the Cenozoic. For instance, in the Miocene, huge amounts of lava (called the Columbia River basalts) erupted in the Pacific Northwest, resulting in the deposition of over two hundred thousand square kilometers of basalt across

Washington, Oregon, and Idaho (Figure 9.5). During the Oligocene, the San Juan volcanic field in Colorado formed from the extensive eruption of andesites. Also around this time, the Cascades, which are composed of a volcanic arc related to subduction of the Juan de Fuca Plate, began to form and are still active today (Figure 9.6). Recent activity in the Cascades includes the 1914 eruption of Lassen Peak in California and the 1980 eruption of Mt. St. Helens in Washington state. Yellowstone National Park, which has been volcanically active since the Pliocene, has been erupting and episodically forming huge plumes of ash that cover much of the western United States in blankets of **tuff**, an ashy volcanic rock. Tuff is useful because it can be dated directly and used to bracket other layers of unknown age.

The extensive orogenic activity in the Cordillera had created a wide belt of highlands by the Oligocene that existed between the West Coast and the Rockies. These highlands became too thick for the crust to remain stable, and by the Miocene, extensive volcanic eruptions were spilling from faults and fissures as the Cordillera began to stretch and thin. This resulted in a vast area composed of north-south oriented mountains and valleys bounded by normal faults. This region is referred to as the Basin and Range province, which lies between the Sierra and Wasatch mountain ranges (the latter found in Utah).

Meanwhile, parts of the divergent boundary between the Pacific and Farallon plates subducted beneath California, which coincided with the formation of the San Andreas Fault Zone, a right-lateral transform plate boundary, during the Oligocene (Figure 9.7). The San Andreas Fault Zone continues to lengthen to the north and south as more and more of the Pacific-Farallon ridge subducts. Currently, there are two triple-junctions of faults that mark the spatial transition from subduction to transform plate movement, and thus the two ends of the San Andreas. One is at the southern end of the Juan de Fuca Plate in northern California, and the other is down in Central America at the northern point of the Cocos Plate. Both the Juan de Fuca and Cocos plates are the remains of what used to be the Farallon Plate. Now, the Pacific Plate takes up most of the Pacific Basin.

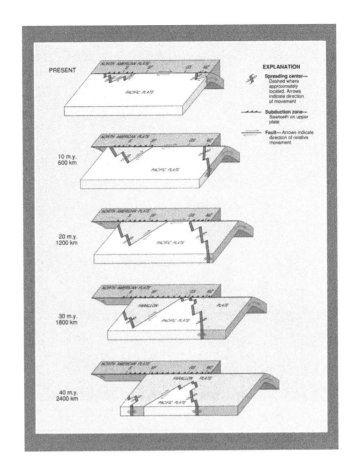

Figure 9.3: Subduction of the East Pacific Rise resulted in the formation of the San Andreas Fault. S) Seattle; SF) San Francisco; GS) Guaymas; MZ) Mazatlan.

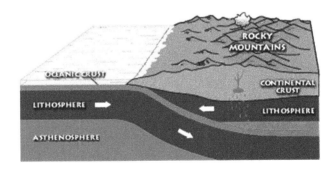

Figure 9.4: Shallow subduction during the Laramide orogeny resulted in the uplift of the Rocky Mountains.

Figure 9.5: Cliff exposure of Columbia River basalt near Portland, OR.

Figure 9.6: A view from the Space Needle shows Mt. Rainier, one of the Cascade volcanoes, in the distance with Seattle, WA, in the foreground.

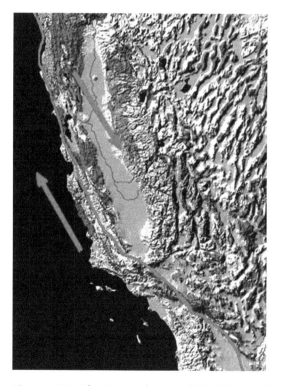

Figure 9.7: The San Andreas Fault (red line) marks the boundary between the Pacific Plate (left) and the North American Plate (right).

PALEOCENE/EOCENE THERMAL MAXIMUM

The climate of the Cenozoic was extremely variable, despite the relative shortness of the era. At the onset of the Cenozoic sixty-five million years ago, the climate was undergoing what was to become the most drastic and abrupt warming periods ever known. The poles were already ice-cap-free, and much of the world basked in tropical to subtropical climates with lots of rain and little difference between winter and summer temperatures. This warming period began in the Late Cretaceous and culminated with a distinct spike of rapid warming 55.8 million years ago during the Early Eocene (about ten million years after the asteroid impact). This drastic period of rapid global warming is now called the **Paleocene/Eocene Thermal Maximum (PETM)**.

During the peak of the PETM, temperatures were on average nearly 7°C warmer across the mid-latitudes than they were at the start of the Cenozoic, with the poles warming by the greatest proportions (Figure 9.8). The oceans also warmed, especially the deep oceans at high latitudes, which warmed up by nearly 8°C. Much of the data we interpret from the PETM comes from the chemistry of oceanic sediments, plankton exoskeletons, and sea shells, which record chemical changes in carbon and oxygen isotopes in seawater associated with climate fluctuations.

The cause of the PETM is unclear but likely linked to a period of increased greenhouse gas concentrations. During this time, India had begun colliding with Eurasia, and there was likely a marked increase in volcanism as the ocean basin closed up between the two during the final stages of subduction prior to continental collision. It is also theorized that large amounts of methane hydrate, usually frozen and sequestered in the sediments atop the ocean floor, were released as the deep-sea temperatures increased. This sudden influx of methane into the atmosphere would have profoundly strengthened the greenhouse effect. The effects of the PETM on the biosphere were profound and will be discussed throughout this chapter in the context of evolutionary trends.

While the Paleocene and Eocene experienced mostly tropical climates, the Oligocene witnessed a period of

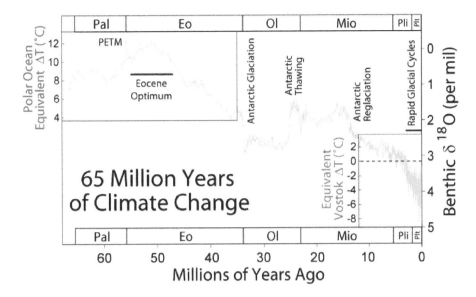

Figure 9.8: A graph showing temperature variations during the Cenozoic. Temperatures are inferred from several methods including measuring oxygen-18 isotope ratios in ice and sediment cores.

cooling and drying that continued almost to this day (the Holocene epoch has been warming up since the last glacial period that peaked about eighteen thousand years ago). The cooling was, in part, caused by the opening of the Drake Passage between Antarctica and South America that kick-started the Antarctic Circumpolar Current, which cooled the southern oceans. Ice sheets on Antarctica and Greenland formed during this period and have been present, to certain extents, ever since. The only significant exception to this cooling trend was a brief warming period about sixteen million years ago during the Miocene. The general cooling period during the second half of the Cenozoic has caused the tropical regions to contract back toward the equator, leaving behind large expanses of open grasslands. Further cooling took place when the Isthmus of Panama closed up around five million years ago, reorganizing oceanic currents and leading to more north-south oriented convection cells such as the Humboldt and Gulf Stream currents that circulate hot and cold water throughout the northern Pacific and Atlantic oceans.

The Pleistocene epoch is, in part, defined by cyclical changes in global temperatures. It appears that these cycles of cold glacial and warm inter-glacial periods have been controlled by variations in the direction and degree of tilt of the Earth's axis, as well as in its distance from the sun. These cyclical climatic fluctuations are referred to as **Milankovitch**

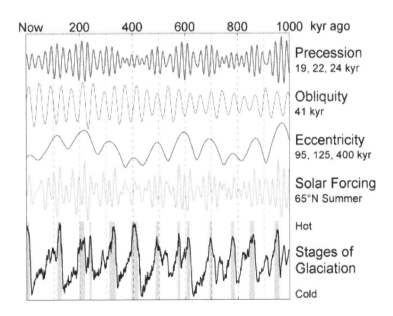

Figure 9.9: Milankovitch Cycles; Precession is the top-like wobble of the Earth's axis. Changes in the angular tilt (obliquity) of the Earth's axis and the eccentricity of the Earth's orbit around the sun. Solar forcing is a measurement of the relative amount of energy coming from the sun.

Cycles, named after the Serbian geophysicist who discovered that these temperature patterns fit well with the cumulative effects of Earth's orbital cycles (Figure 9.9). These orbital cycles make it easier for ice sheets to grow in the Northern Hemisphere at particular times. The last four to five glacial cycles have occurred on average about 100,000 years apart. Prior to this, the cycles were more variable, sometimes as short as 40,000 years. During these glacial periods, ice sheets and alpine glaciers (mountain glaciers) grew and sea level fell. Much of the northern part of the United States, especially in the areas of the Great Lakes and the eastern seaboard, was buried beneath thick lobes of ice that advanced south from Canadian ice sheets. The Great Lakes occupy broad, deep valleys that were gouged out by lobes of ice over several cycles of glaciation.

CENOZOIC INVERTEBRATES

MARINE

During the K-T mass extinction, many different types of marine plankton reduced in numbers, especially planktonic foraminifera, which became primarily benthic afterwards. Most of the plankton genera that did survive went on to diversify during the Cenozoic.

Coccolithophores, the marine phytoplankton that produce tiny, hubcap-shaped platelets (coccoliths) made of calcite, made it through the K-T boundary, although over 60 percent of their families went extinct. As the oceans warmed during the Paleocene and Eocene and atmospheric concentrations of CO_2 increased, coccolithophores thrived in the open oceans and took advantage of the increased CO_2, producing large amounts of coccoliths (Figure 9.10). As the climate cooled after the PETM, the temperatures in the oceans became more latitude-dependent and ocean temperatures in the mid-to-high latitudes decreased. Many new coccolithophores adapted to these

Figure 9.10: The Cretaceous White Cliffs of Dover in southeastern England are composed almost entirely of coccoliths.

cooler temperatures, and today, they mostly occupy cool, sub-polar waters.

Diatoms, like coccolithophores, are marine phytoplankton, but they create shells made of silica instead of calcite (Figure 9.11). They survived the K-T extinction and diversified toward the end of the PETM. They increased in numbers during the Miocene, when volcanic activity had increased. The fallout of volcanic ash into the oceans supplied huge amounts of dissolved silica to the water, giving diatoms plenty of material from which to construct their shells. Despite this, many diatoms have since evolved to be more conservative with their use of silica; their tests are now thinner, more porous, and less ornate than they were earlier in the Cenozoic. Diatom species that are more adapted to cool waters evolved and diversified during and after the Miocene when the Earth was in the process of cooling down. The Miocene Monterey Formation of coastal California is an

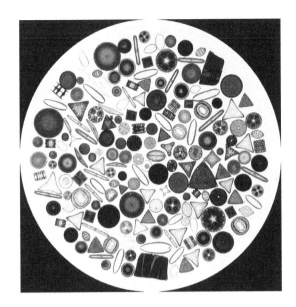

Figure 9.11: An assortment of diatoms displaying intricate silica exoskeletons.

important resource of oil and gas, with much of the fossil fuels derived from diatoms. In addition, the fine-grained nature of these deposits has allowed for the excellent preservation of fossil seaweed, whales, dolphins, and a variety of marine invertebrates, such as crabs.

Foraminifera, as mentioned earlier, suffered losses of many planktonic species during the K-T extinction, but the benthic species fared a little better. Cenozoic foraminifera were generally smaller than their Mesozoic ancestors, although some benthic species evolved to be rather large during the PETM.

Dinoflagellates, if you remember, are single-celled protists that compose a diverse group of poisonous plankton that create cellulose tests to protect themselves when they go dormant in the winter. A particular dinoflagellate called *Apectodinium* flourished in the warm surface waters of the PETM, fed by large amounts of nutrients being pumped into the oceans from rivers during this period of increased precipitation.

Reef systems underwent some major changes during the K-T mass extinction. One major change was the extinction of the rudists, the prolific Cretaceous bivalves that crowded out the corals and took up most of the space in reefs. At the beginning of the Cenozoic, many corals that had been used to living on the periphery of reefs began to move back in and rapidly diversify. These reef-building corals, called scleractinians, had been generally successful throughout the Cenozoic and were likely plentiful before and during the PETM, when warm oceanic conditions allowed for the expansion of reef systems into mid-latitudes. Three major events affected corals during the late Cenozoic. First, when the Isthmus of Panama closed up around five million years ago, coral communities nearer the Pacific were isolated from the Caribbean groups, creating a barrier to genetic transfer. Second, the Pleistocene Ice Age resulted in drastic drops in sea levels that exposed and killed off many reef communities. Third, recent increases in atmospheric CO_2 are resulting in ocean acidification, which promotes the dissolution of calcite, the main structural component of coral and many other invertebrates.

Bryozoans continued to thrive throughout the Cenozoic and are still with us today. While the descendants of lacy bryozoans of the Paleozoic and Mesozoic are still around, many new forms of encrusting bryozoans have evolved to cover the surfaces of shells, rocks and other substrates as opposed to forming delicate branches (Figure 9.12). The encrusting forms are sometimes problematic, as they coat the pilings of piers and docks, and some freshwater species clog up water intake pipes used for public or industrial purposes.

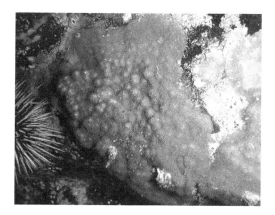

Figure 9.12: A modern encrusting bryozoan.

Figure 9.13: Fossils of an Eocene gastropod, *Clavilithes laevigatus.*

Mollusks were able to survive the K-T extinction, but with some exceptions, such as the rudist bivalves. Many other bivalves, such as clams and mussels, thrived throughout most of the Cenozoic and are still very common today. Gastropods diversified early in the Cenozoic and have been successful to this day, making up 80 percent of all mollusks and being the only members of the group to move onto land (Figure 9.13). Their feeding habits include suspension feeders, detritivores, scavengers, and even carnivores. A major radiation of sea snails took place in the Late Cretaceous and Paleocene, and of land snails during the Eocene. Most modern families of marine, freshwater, and land snails date to the Oligocene and Miocene. Shelled cephalopods such as ammonites dominated the Mesozoic, but they did not make it through the K-T mass extinction. The only shelled cephalopods to do so were the nautiloids, which are still living today; otherwise, the rest of the Cenozoic cephalopods have lacked shells (Figure 9.14). Octopuses and squid were, and are, the most common cephalopods of the Cenozoic. The fossil record of shell-less cephalopods is poor; some body fossils of octopuses have been found in Jurassic rocks, but the majority of fossil evidence of octopuses and squid comes from Cenozoic rocks.

Echinoids such as starfish and sea urchins first evolved in the Paleozoic and have been important members of marine communities ever since. Common echinoids of the Cenozoic include starfish, sea urchins, and sand dollars (Figure 9.15). Sand dollars are the most recent of these three to evolve; their ancestors can be traced back to Jurassic sea urchins. By the start of the Cenozoic, sand dollars had taken on a mostly familiar form, having evolved from biscuit-shaped ancestors. Sand dollars such as the Late Paleocene *Togocyamus* show this relict morphology.

Crabs, one of the many crustaceans to survive the K-T boundary, had, in the Late Cretaceous, evolved an enlarged claw designed for crushing or peeling open the shells of aquatic snails. This is an adaptation that has sprung up within several crab lineages and is common today. Several new types of burrowing shrimp evolved in the Cretaceous and continue to exist. They are particularly harmful to oyster farms, where they stir the mud, causing the oysters to suffocate or sink too far into the substrate. Many more genera of shrimp continue to live pelagic or mobile lifestyles.

TERRESTRIAL INVERTEBRATES

By the beginning of the Cenozoic, most insects had taken on familiar forms; those that had already evolved include cockroaches, beetles, snails, slugs, flies, grasshoppers, wasps, earwigs, termites, moths, and bees. In addition, spiders and other arachnids (scorpions and mites) had taken on a modern appearance. The fossil record of insects is always sparse, and the Cenozoic is no exception. A more modern approach to studying the behavior of some insects involves

Figure 9.14: Examples of modern cephalopods.

the analysis of insect-induced damage to fossil leaves. Many fossil leaves from the PETM indicate a greater diversity of leaf-eating insects (such as beetles belonging to *Cephaloleia*) in warm, subtropical to tropical regions. Despite this diversification of arboreal insects, trace fossils of burrowing insects appeared to have become thinner on average, indicating a reduction in size of ground-dwelling insects. Toward the end of the PETM, the typical form of leaf damage changes from mining (Figure 9.16), where trails are eaten through leaves, to galling, which is more localized and causes leaves to form bumps.

One new insect to thrive during the Cenozoic was the butterfly, which evolved from moths (and more indirectly, caddisflies) sometime during the Late Mesozoic or Early Cenozoic (Figure 9.17). In fact, the earliest direct fossil of a butterfly comes from the Eocene and is about fifty million years old. Some researchers suggest the already-specialized nature of this fossil indicates that the original butterflies may date as far back as the Cretaceous. The transition from moths to butterflies is a subtle one, and there are scientists today that consider butterflies as "gaudy moths." Several differences can be observed between moths and butterflies: moths are mainly nocturnal, while butterflies are mainly diurnal;

Figure 9.15: Living sand dollars at the Monterey Bay Aquarium, CA.

Figure 9.16: A fly larva in the process of mining a leaf for nutrients.

Figure 9.17: A peacock butterfly showing a bright array color on its wings.

Figure 9.18: The undergrowth of this jungle in the Sierra Madre in Mexico displays large leaves designed to gather the maximum amount of sunlight that filters in through the dense canopy above.

the wings of moths fold down and back, while those of butterflies only fold upwards; and butterflies have forewings and back wings that are not attached to each other like the wings of moths.

PLANTS

By the start of the Cenozoic, angiosperms were on their way to becoming the most dominant land plant. The fossil record of angiosperms includes both direct fossils of the plant tissue as well as fossil pollen, which are often found mixed in with ancient sediments. One of the most direct lines of evidence indicating the PETM is the fossil record of plants, which are especially helpful in determining ancient climates (Figure 9.18). The Paleocene was an epoch of warming that culminated in the Eocene. Tropical and subtropical forests spread north and south from the equator, reaching sub-polar regions such as the Bering Strait, indicating a much warmer climate there than today.

Leaves serve several functions, such as the capturing of light and absorption of carbon dioxide for photosynthesis. To avoid drying out, leaves are coated in a waxy coating called cutin. Small pores called stomata allow carbon dioxide to enter the leaf to be used for photosynthesis. When the stomata open for the intake of carbon dioxide, water may be lost from the inside of the leaf. Therefore, leaves must be designed to maximize light and gas intake without losing too much water in the process. Depending on the climate, these variables differ, and the leaves of plants have various shapes that allow them to get the most benefit from these conditions. The **physiognomy** (sizes and shapes) of leaves tell us a lot about the climates in which they grew (Figure 9.19).

Leaf length is related to the mean annual temperature (MAT) of the location in which it grew. The longer the leaf, the higher the MAT. *Leaf area* is related to both temperature and precipitation. Leaves with large areas tend to grow in warm, humid climates. Leaf area is also related to the amount of available sunlight; leaves that grow in the shady undergrowth of tropical forests tend to be the largest, as they must maximize light intake. *Leaf width* is related to precipitation, with wider leaves growing in areas that receive more rain. Leaves in these wet climates also tend to have a pointy drip-tip to allow for efficient water drainage off the surface of the leaf. Pine needles are some of the thinnest leaves relative to their length and are highly adapted to variable climates that get cold in the winter. The bunching pattern of pine leaves decreases water loss while increasing surface area (through many thin leaves as opposed to less numerous wide leaves) for the collection of light and transport of gases. The correlation of leaf shape to MAT is so strong that several formulas have been established to determine past temperatures using the relative percentages of rounded to jagged leaves found at a particular fossil site or within close but separate rocks of similar ages.

Besides the shape and area of the leaf, another indicator of climate is the smoothness of the leaf margins. Leaves with smooth edges (called "*entire*" margins) are typical of warmer climates, while those with "teeth" along the edges (called **serrate** leaves) indicate cooler climates. Leaves with serrated margins have increased sap flow, allowing leaves to begin photosynthesis earlier in the spring (when it is still relatively cold) than those in warmer climates. Serrated leaves also tend to lose water more easily than smooth-margined leaves (they typically have more stomata), which is why serrated leaves are limited to more temperate climates where they are less likely to dry out due to the heat.

The PETM is recorded in the fossil record of plants as an increase in the amount of rounded, smooth-margined fossil leaves across most of the terrestrial world, including places today that are dominated by glaciers and permafrost. Also abundant in the mid-latitudes were ferns and palms. By the Oligocene, when climates began to cool, leaf fossils begin to appear more serrated and generally smaller in size throughout the mid- and high latitudes. Grasslands began to replace the mixed deciduous and evergreen forests during the Late Eocene, and the tropical and subtropical forests receded back to the equatorial regions during the Oligocene. The sudden increase in grasslands across the mid-latitudes of North America coincides with the evolution of ruminants, herbivorous mammals with grinding teeth designed for a diet composed primarily of grasses.

The cooling period following the PETM was accompanied by a decrease in precipitation across what are today the desert belts around 30

Figure 9.19: Comparison of a rounded modern, and fossil, leaf (left) and a serrated modern, and fossil, leaf (right). Fossil leaves become more serrated in the fossil record during the cooling following the PETM.

degrees north and south of the equator. Many plants had to adapt to these drier conditions, such as conifers and succulents. While conifers have adapted to cooler environments by growing long, thin, evergreen needles, succulents, which are angiosperms, have evolved to withstand long, dry periods with little to no rain by efficiently absorbing and retaining what little water there is available. One way they do this is by growing extensive, shallow root systems designed for maximum absorption of water from the dew-moistened desert topsoil. Another adaptation is that they open their stomata at night instead of the day so that water loss is minimal during the cooler nighttime temperatures. Furthermore, succulents are especially fleshy to allow for maximum water storage.

The fossil record of succulents, especially cacti, is extremely poor. It may be that they didn't grow in large enough numbers to occupy much of the fossil record, or perhaps the dry environments in which they grow do not favor fossilization. The earliest evidence of succulents are pollen grains from the Cretaceous, and ground sloth dung fossils from the Early Cenozoic also include fragments of what may be cactus needles. Otherwise, the fossil record of succulents is still too poor to establish any confident phylogenetic relationships of these highly specialized plants. One fairly certain relationship is that they evolved from earlier angiosperms. Another interesting fact is that all endemic cacti are found in the Americas and have only been introduced to other parts of the world as invasive species.

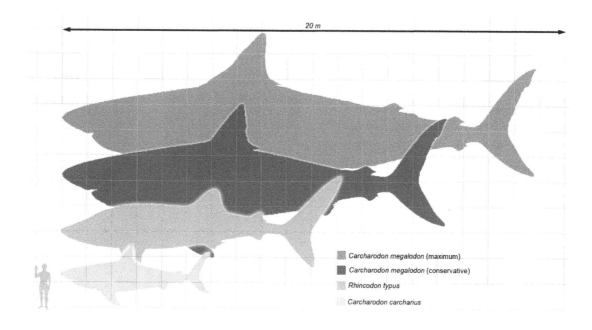

Figure 9.20: Comparison of *Carcharodon megalodon* with other large sharks. The modern Great White shark is shown at bottom next to an unfortunate human (waving goodbye) for scale.

CENOZOIC VERTEBRATES

FISH

Fish in the Cenozoic had already become dominated by teleosts (ray-finned bony fish) with a smaller, but still significant, group of cartilaginous fish such as sharks, rays, and skates. Teleosts live in freshwater as well as the oceans, and their adaptive radiation in the Mesozoic and Cenozoic is manifested in the great variety of extant species.

Sharks became the biggest fish of the Cenozoic, rivaling the Jurassic *Leedsichthys* in size. The *Carcharodon* genus of sharks, which include today's Great White, reached maximum lengths of up to sixty feet, such as the *Carcharodon megalodon* that lived from the Oligocene to Pleistocene (Figure 9.20). *Megalodon* was a more robust, and much larger, version of the Great White and was the apex predator of the seas during the Late Cenozoic (Figure 9.21). It fed on prey such as whales, porpoises, and giant turtles.

Figure 9.21: A three-inch-long tooth of *Carcharodon megalodon*, ~25 million years old.

AMPHIBIANS

Most amphibians throughout the Cenozoic have been represented by the lissamphibians, those that are modern in appearance. As mentioned in the previous chapter, the origins of lissamphibians are sketchy and may or may not lie in the Temnospondyls. It is also possible that only some of the modern amphibians were derived from the Temnospondyls and the rest from other groups.

There are three orders of lissamphibians today (Figure 9.22). **Anurans** (meaning "without tail") compose the superorder Salientia and includes frogs and toads. This group has an especially elongate ankle bone, giving them the ability to jump. The anurans can be traced in the fossil record to as far back as the Jurassic, although early proto-frogs, such as *Triadobatrachus*, date to the Early Triassic. The **Caudata** are composed of the Salamandridae family, which are essentially salamanders, including their subgroup, newts. This group is highly diverse, from those that live aquatic lives to those that live on land during their adulthood. They range in length from a few centimeters to several feet. A unique adaptation of salamanders is their ability to regenerate lost limbs. Some species are also highly poisonous. Salamanders generally have elongate bodies and tails, with sprawling limbs that are generally small. They sometimes lack a pair of limbs, such as the siren, an aquatic amphibian that has lost its hind limbs.

The third order of lissamphibians are the lesser-known **Gymnophiona**, also called Caecilians, which are limbless, burrowing creatures that resemble large earthworms, growing to lengths of over five feet. They have very poor eyesight, as they spend most of their time underground. Their bodies are highly specialized for digging into mud and other soft sediments.

REPTILES

Reptiles were hit rather hard by the K-T mass extinction. Lost were the non-avian dinosaurs as well as pterosaurs, mosasaurs, and plesiosaurs. The only marine reptile to survive was the sea turtle, which is still around today. Cenozoic reptiles were overall much smaller than their Mesozoic ancestors. The largest terrestrial reptiles to survive the K-T extinction were turtles, crocodiles, and the *Champsosaurus*, a reptile superficially similar in appearance to crocodiles that went extinct during the Eocene (Figure 9.23).

Figure 9.22: Examples of modern amphibians: a frog (top), an anuran; a salamander (middle), a member of Caudata; and a caecilian, a member of Gymnophiona.

Smaller reptiles were better able to survive the mass extinction, probably because when food is limited, smaller organisms are less affected. In addition, they could hide in burrows or under rocks and logs for shelter. About ninety-five percent of modern reptiles belong to the order Squamata, which includes nearly ten thousand species of lizards and snakes. The other five percent includes, among others, tuatara, turtles and tortoises. Squamates, especially snakes, diversified during the Cenozoic. Snakes appear in the Cretaceous and may have evolved from

Figure 9.23: A Cretaceous fossil of a *Champsosaurus*.

burrowing lizards, which would benefit from a limbless, streamlined body. Another theory is that snakes evolved instead from marine reptiles that recolonized land after their limbs had evolved away. One particularly amazing fossil from India is of a twelve-foot Cretaceous snake eating a newborn sauropod. Recent studies suggest that sauropods may have adapted to the nest-raiding predatory methods of snakes by growing rapidly in their early years. Another uniquely adapted lizard is the *gecko*, whose fossil record can be traced back to the Cretaceous (Figure 9.24). Geckos have specialized padded feet that adhere to most surfaces, even glass. The exact nature of how the gecko's feet create such adhesion is still under study, but it appears that the reasons lie on the atomic level. Overall, terrestrial reptiles have recovered quite well from the mass extinction event, although during the Cenozoic, they remained relatively small compared to those from the Mesozoic.

CENOZOIC BIRDS

Numerous groups of both flying and flightless birds went extinct at the end of the Cretaceous, including many of the archaic carnivorous species. Additionally, some of the best fliers, such as the Enantiornithes, as well as some of the most agile flightless waterfowl, like the Hesperornithiformes, went extinct during the K-T event.

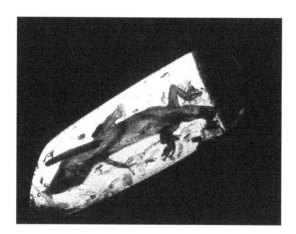

Figure 9.24: A rare Oligocene gecko fossil trapped in amber.

The Neornithes subclass, which is composed of today's species and their direct ancestors, were the only group of birds (technically, avian dinosaurs) known to survive the K-T mass extinction event. The fossil record indicates that the majority of early Neornithes were flying species. Nearing the end of the Cretaceous, however, the Enantiornithes had surpassed Neornithes as the dominant group of flying birds, and the Neornithes became represented by a growing proportion of flightless varieties. By the *very* end of the Cretaceous, however, Neornithes had suddenly diversified into a new spectrum of flying varieties, and it was *them* and their ground-dwelling brethren that survived the extinction event.

Of the Paleogene flightless birds, one of the most famous is the genus *Gastornis*, which went extinct during the Eocene. Members of *Gastornis*, including the well-known bird informally referred to as "Diatryma," were on average about two meters tall and had long, powerful legs, giant beaks, and robust vertebrae (Figure 9.25). Their wings, however, were vestigial, the leftover remains of their flying ancestors. Therefore, species of the genus *Gastornis* were built much like their theropod predecessors.

There is some contention about whether *Gastornis* was a group of carnivores, as traditionally thought, or rather herbivores, as new research suggests. In 2009, a landslide in Washington State exposed the surfaces of Eocene

Figure 9.25: Size comparisons of "terror birds" from the Early Cenozoic.

mudstone beds, which contains numerous footprints of what may be Diatryma. Surprisingly, the footprints do not show impressions of large claws. Furthermore, the upper beak of Diatryma does not have the typical hooked tip diagnostic of carnivorous raptors (Figure 9.26). The lack of sharp claws and a hooked beak may indicate it was unlikely that Diatryma, and the rest of *Gastornis* for that matter, were carnivores. Instead, they may have fed on vegetation, seeds, and fruits. *Gastornis* belongs to the order Anseriformes, which also includes many of today's waterfowl, such as ducks, geese, and swans. This means that in the context of Linnaean classification, Diatryma is more related to water-fowl than to the ostrich and emu, which are members of a completely different order (Struthioniformes).

Figure 9.26: Skull of *Gastornis* at the National Museum of Natural History in Washington, DC.

Other orders of birds did, however, include large, flightless carnivores. The order *Cariamae*, which includes today's falcons, kestrels, parrots, and songbirds, include, according to some theories, the extinct Phorusrhacidae family. The Phorusrhacids ("terror birds") showed up in the Early Paleocene, lived throughout most of the Cenozoic, and went extinct during the Pleistocene. Phorusrhacids were, by far, the most formidable terrestrial predators of the Early Cenozoic. Unlike the *Gastornis* genus, members of the Phorusrhacidae family were well-equipped to hunt; they had a hooked beak, sharp talons, and heights of up to ten feet. Estimates indicate that some species of Phorusrhacids could run up to thirty miles per hour.

Examples of the Phorusrhacidae family, such as *Titanis*, *Kelenken*, and *Brontornis*, have unusually large and pronounced curved tips to their beaks. These curved tips may have been used like a pickaxe, wherein the bird would strike down and break the spine or cranium of its prey. Another method of incapacitating prey may have been to pick them up and slam them against trees or the ground, like some raptors do today. These methods would likely have been employed to kill larger-sized prey and allow the bird to use its beak to tear large bites of flesh off. They likely fed on smaller prey as well, which would have been easier since they could be gulped down in one bite.

While the Gastornis and Phorusrhacidae families were still roaming the Earth, *today's* members of the Neornithes subclass had already begun to look modern in appearance. Extant Neornithes are composed of distinct lineages that likely diverged in the Late Cretaceous. By far, the most numerous of the two are the Neognaths, which compose most of today's birds. Neognaths such as owls, hawks, penguins, vultures, chickens, and turkeys evolved in the Early Cenozoic, and by the Miocene, modern songbirds had evolved. The other lineage of Neornithes is the Paleognaths, today's flightless birds, which include ostriches, emus, and kiwis, among others. Paleognaths evolved by the Paleocene and may have roots in the Cretaceous. By the Miocene, the direct ancestors of today's songbirds (members of the Neognaths) had evolved, and by then the variety of birds on Earth was very similar to that of today.

CENOZOIC MAMMALS

During the Mesozoic, mammals remained small, mostly rodent-sized, likely due to the strong foothold that dinosaurs had established in most ecological niches reserved for terrestrial vertebrates. It is unlikely that competition for food was the reason for the stunted evolution of Mesozoic mammals; many were likely nocturnal and would have had their fair share of insects to eat during the night. Rather, it is thought by many scientists that predation from dinosaurs was the main reason why mammals remained small. Being small was advantageous because it allowed mammals to scurry under rocks or dense foliage to escape carnivorous reptiles.

It wasn't until the non-avian dinosaurs went extinct that mammals were able to emerge from their sheltered nooks and establish themselves in a wide variety of freshly-vacated niches. It is no surprise, then, that by the Early Cenozoic, a significant period of mammalian adaptive radiation took place. Much of this diversity was established in the **placental** group of mammals, and today, they compose the majority of mammalian orders.

Monotremes, on the other hand, never reached the level of success of their mammalian counterparts. Today, extant monotremes, of which there are only five species, live in Australia and New Guinea. A single tooth of what some scientists consider an Early Cenozoic monotreme has also been found in Argentina, suggesting monotremes had a larger range across Gondwana than just the area that tore off and became Australia. The earliest monotremes, which evolved in the Cretaceous, were very similar to today's platypus both in size and in shape, including a "billed" snout and sprawling posture.

The duck-billed platypus is an extant species of monotreme that has developed some strange adaptations (Figure 9.27). They are primarily nocturnal and live in burrows along the shores of freshwater lakes and rivers. The platypus is well adapted for swimming, having webbed feet and the ability to remain underwater for up to

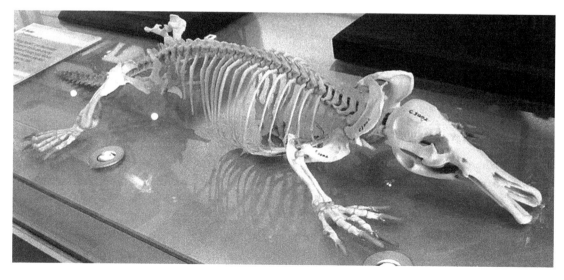

Figure 9.27: Skeleton of a platypus, a modern monotreme.

ten minutes. This is useful, as they are insectivores and much of their food, such as insect larvae and worms, live in the mud beneath the water. While the platypus swims and digs through the mud with its bill, its eyes and nostrils are closed tight. It finds its food instead using specialized electroreceptors on its bill. Male platypuses are also poisonous and use a small spur on their hind limbs to inject deadly poison into other platypuses during competition for mating. Of all the mammals, this is perhaps one of the most dire methods of sexual competition.

The rest of today's monotremes consist of four species of echidnas, also known as spiny anteaters. Like a porcupine, they have spines covering their entire body, and as a defense mechanism, they quickly dig burrows, or if worse comes to worst, curl up into a spiny ball. As their name implies, echidnas are insectivores with a diet of ants, worms, and various other insects. They have no teeth but have an elongate, sticky tongue, narrow snout, and sharp claws, allowing them to easily dig into the ground or rotting logs to collect insects. Like the platypus, echidnas have electroreceptors on their snout, although not as sensitive. Results of genetic research indicate that echidnas diverged from the platypus lineage sometime during the Mid-Cenozoic, likely due to adaptations toward a drier, more terrestrial environment chosen by basal echidnas.

Marsupials, although outcompeted by placentals during the Cenozoic, nevertheless are much more diverse than the five species composing the monotremes. In fact, North America during the Cretaceous had a greater amount of marsupials than placentals. However, the K-T mass extinction was particularly tough on North American marsupials, and most genera there disappeared by the start of the Cenozoic. As North American placentals diversified in the Paleogene, they began to outcompete the surviving marsupials and caused their diversity to shrink even more. Today, the only extant species of marsupials in North America is the opossum. South America also had a rather diverse population of marsupials that lasted much longer into the Cenozoic than they did in North America. However, when the Isthmus of Panama connected in the Early Pliocene, North American placentals migrated to South America and caused marsupials to decline in numbers there as well. Extant marsupials in South America consist of a slightly more diverse group of opossums than present in North America. Today, the only place where marsupials remain successful and diverse is Australia. Surprisingly, the abundance of Mesozoic marsupials in what is now North America suggests that they originated there and migrated southward into Gondwana prior to the complete breakup of Pangaea.

Marsupials at the start of the Cenozoic were not unlike today's opossums, which are considered to be an example of a "**living fossil**." A Paleocene marsupial fossil from Bolivia called *Pucadelphys* was a burrowing, opossum-like creature that likely had diverse diets like opossums of today: insects, fruits, and small vertebrates. By the Late Paleocene, these opossum-like basal marsupials diversified into a range of sizes, from the mouse-sized *Minusculodelphis* to the larger, carnivorous *Eobrasilia*. A more rodent-like group of marsupials (based on their small size and large front teeth) called polydolopids also evolved in the Paleocene and became an important and diverse group that lasted through the PETM and finally went extinct during the Oligocene.

The carnivorous lineage of marsupials, first seen in examples like *Eobrasilia*, lasted most of the Cenozoic. *Sparassodonta* (Figure 9.28) is a now-extinct order of South American carnivorous mammals (thought by many to be marsupials) that evolved in the Paleocene, peaked in the Miocene, and went extinct by the Pliocene. Together, with the Phorusrhacids (terror birds), they were the apex predators of the Early Cenozoic and show many convergent characteristics with carnivorous

Figure 9.28: Fossil of *Sparassodonta*, an early marsupial mammal, Natural History Museum of New York.

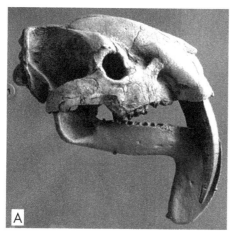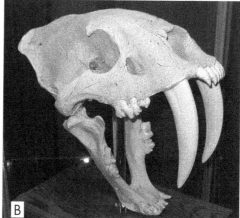

Figure 9.29: Comparison of the marsupial *Thylacosmilus* (left) with the placental *Smilodon* (right) of similar age.

placentals. For example, the Miocene-Pliocene *thylacosmilids* were built sort of like a sabre-toothed cat (but weren't related). They had very large canines and an odd lower mandible that bent downwards, almost like a holster for the large sabre teeth to fit in. They reached lengths of up to six feet and were adept and successful hunters, at least until their place at the top of the food chain was supplanted by placentals such as *Smilodon* after North and South America connected in the Pliocene (Figure 9.29).

During the Eocene, it is theorized that a group belonging to the *Microbiotheria* family, which are small, tree-dwelling marsupials, migrated from South America, across Antarctica, and into Australia. This theory is supported by molecular studies that indicate the marsupials in Australia are descendants of South American ancestors. This means that the variety of marsupials seen today in Australia, including kangaroos, koalas, wombats, Tasmanian devils, and the recently extinct Tasmanian wolf, may have been derived from a common marsupial ancestor that migrated from South America in the Eocene (Figure 9.30). Without competition from placental mammals (except some rodents since the Pliocene), marsupials in Australia have diversified and came to fill a plethora of ecological niches over the past fifty million years. Marsupials in South America would likely still be diverse today had the Isthmus of Panama not connected the Americas.

Within ten to fifteen million years of the K-T mass extinction, placental mammals had diversified into most of today's modern orders, including bats and whales. Perhaps the short life spans and ability to produce greater amounts of offspring allowed placentals to diversify so well in such a short period of time. Several of the more significant clades of placental mammals are discussed in the following pages.

Multituberculates were one of the longest-lasting orders of mammals; these rodent-like creatures first appeared in the Late Jurassic or Early Cretaceous but went extinct during the Oligocene, when they were outcompeted by placental (true) rodents. Had they survived, they would compose a fourth major group of mammals, in addition to monotremes, marsupials, and placentals. Their name is derived from the many cusps on their teeth. Based on the shape of their pelvis, it appears that multituberculates had similar birthing habits to marsupials, wherein the young is born underdeveloped.

Multituberculates remained generally small, ranging in sizes between mice and beavers. Their heads were large and wide, with downward-sloping snouts, cusped molars, blade-like premolars, and in some species, sharp canines. Most multituberculates had robust cheekbones upon which strong chewing muscles attached. Their diets are not fully understood, but as their dentition is similar in many respects to that of rodents (a large gap between the back and front teeth), they, too, likely ate seeds and nuts. The earliest known multituberculate is the Cretaceous *Catopsbaatar*, which looked similar to a large rat (Figure 9.31). Some Paleocene species, such as

Figure 9.30: Examples of modern marsupials.

the Asian *Lambdopsalis*, lived in burrows, while others, like the North American *Ptilodus*, were more squirrel-like and climbed trees to access fruits, nuts, and insects. It had a prehensile tail, a convergent adaptation seen in many other arboreal mammals.

There were other groups of Early Cenozoic giant mammals, such as the *Pantodont* suborder, which included some of the earliest placentals. Early pantodonts were small and likely arboreal. Later examples, such as the Oligocene *Coryphodon*, were larger and had a terrestrial to semi-aquatic lifestyle, like modern hippos.

The order Dinocerata also included some

Figure 9.31: Illustration of the multituberculate *Catopsbaatar*.

rather large genera, such as the Eocene *Uintatherium*, which was shaped somewhat like a thirteen-foot-long, two-ton rhino with large fangs and an odd-shaped skull that was concave above the snout (Figure 9.32).

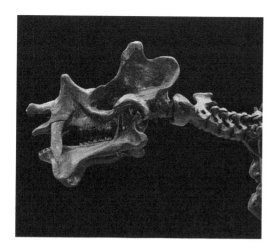

Figure 9.32: Skull of *Dinoceras mirabile*, a Middle Eocene *Uintatherium*.

RODENTS AND LAGOMORPHS

Like the extinct multituberculates, the body plan and lifestyle of most rodents have remained more or less similar to mammals from the Cretaceous. Rodents appear in the Paleocene, and their arrival coincides with the decline of the multituberculates. One way that rodents had an advantage over the multituberculates is that they possess teeth that grow continuously, preventing the enamel from getting too worn down from gnawing food. In addition, the enamel on the front of their lengthy pair of incisors is thicker than the back, making it more resistant to abrasion than the back of the tooth; this keeps the tooth sharp and chisel-shaped as it wears. Rodents are today the largest group of mammals, having over fifteen hundred modern species. Modern rodents include beavers, porcupines, woodchucks, chipmunks, squirrels, mice, rats, guinea pigs, hamsters, prairie dogs, gophers, lemmings, capybaras, and many more.

Early rodents likely branched off from a small group of Paleocene mammals called *anagalids*, guinea pig–like creatures that may be the common ancestor of rodents, *lagomorphs* (e.g., rabbits and hares), and *insectivores*.

Some extinct rodent genera (mostly from South America) reached massive sizes. The Late Miocene-Early Pliocene *Phoberomys pattersoni* from Venezuela reached lengths of ten feet, not including their five-foot tail. This giant rodent lived in marshy environments, such as lagoons and wetlands. Another slightly more recent South American giant from the Pliocene named *Josephoartigasia monesi* is, as of now, the largest known rodent ever to exist. It weighed up to fifteen hundred pounds, was built like a large bull, and had a general resemblance to today's capybara, the largest *modern* rodent. Its closest living relative, however, is the Amazonian *pacarana*. *Josephoartigasia* lived in similar shoreline environments to those of *Phoberomys*. Like their ancestors, today's giant rodents, like the pig-sized capybara, still reside in South America. It should be noted that there are some exceptionally large rats (up to sixteen inches long) living in the subways of New York as well. One of the largest extinct North American rodents was the Pleistocene *Castoroides*, a giant, bear-sized beaver (Figure 9.33). There is one fossil rodent species of note from the Late Miocene called *Ceratogaulus*; it was not very large but had a different, unique trait. It was the only known example of a rodent possessing a pair of defensive horns on its snout (Figure 9.34).

Figure 9.33: Fossil of the giant beaver *Castoroides*.

Figure 9.34: Illustration of *Ceratogaulus*.

Most rodents of today are small, like the Muridae family, which includes mice, rats, gerbils, hamsters, and other small rodents. They have a poor fossil record, and so far, it appears they radiated in the Early Miocene and then again in the Holocene, when ships brought many species to new lands. Murids are frequent breeders and give birth to large litters; this and their short lifespan allow them to adapt quickly, so it is no surprise they are very common today.

Lagomorphs (rabbits, hares, and picas) also likely diverged from the anagalids in the Paleocene or Eocene. Lagomorphs differ from rodents in that they have four incisors on the upper jaw, generally shorter tails, and powerful, elongated hind limbs, allowing them to leap and absorb the landing on their smaller front limbs. They reached their

Figure 9.35: *Nuralagus,* the largest lagomorph known in the fossil record.

peak of diversity in Mid- to Late Cenozoic and are now in decline. Like the rodents, lagomorphs also achieved some large sizes, such as the Pliocene *Nuralagus*, which reached lengths of about four feet long (Figure 9.35). Unlike modern rabbits, it had short ears and didn't hop; its back legs were more proportional to its front, giving it the overall shape of a large marmot. Rabbits and hares likely developed long hind feet for several reasons, including better balance so they can stand upright and for easier digging of burrows.

The reason they hop is because their hind feet are too large to "walk" with.

UNGULATES

Ungulates, which include herbivorous hoofed placentals, are divided into two types based on the number of their toes; **Artiodactyls** have an even number of toes (two or four), and **Perissodactyls** have an odd number of toes (usually one or three, but in earlier species, up to five). The general trend of ungulate evolution throughout

Figure 9.36: Illustration of Chalicotheres.

the Cenozoic was toward a larger size, fewer number of toes, and an increasing reliance on the middle toe(s) for supporting most of the weight.

To no surprise, the earliest ungulates were perissodactyls, as the first hoofed animals still retained all five digits characteristic of most vertebrates. Perissodactyl ungulates can be traced back to the Paleocene, and the first artiodactyls, having first lost their fifth toe to vestigial oblivion, showed up soon after in the Eocene.

An interesting, now-extinct family of perissodactyls was the knuckle-walking *Chalicotheres*, which were about the size of horses but quite different in shape (Figure 9.36). These Eocene-Pliocene ungulates had robust front limbs with large claws that were likely used to pull down tree limbs to access leaves. Their hind limbs, however, were strong but short, giving them a gait similar to gorillas. Another extinct family was the Eocene-Oligocene *Titanotheres*, which started out the size and general shape of rhinos but evolved into progressively taller forms, up to eight feet. Some species had distinctive sets of bony horns near their noses that were likely an adaptation for sexual competition between males. They were mostly browsers (grass-eaters) and lived in more open, grassy areas than many of the Chalicotheres.

Today's perissodactyls include horses, tapirs, and rhinoceroses. The earliest direct ancestors of horses include the North American, Paleocene *Eohippus*, which was the size and general shape of a dog. Its front feet had four toes and its back feet had three. It was slender, like a deer, and had teeth adapted for chewing leafy vegetation. *Eohippus* (*Hyracotherium*) is a popular member of a group of various early horses that wandered the undergrowth of forests. Possibly, this environment favored a small size, as they could wander more easily between closely-growing trees and other vegetation. In the Eocene, larger genera such as *Mesohippus* showed up. *Mesohippus* had only three toes in the front; the middle was by far the largest and supported the majority of the horse's weight. By the Miocene, at least a dozen species of horses were living in North America and ranged in size from a couple feet, such as the grazing (shrub-eating) *Nannippus*, to nearly the size of modern horses, like the browsing *Dinohippus*. By the Late Miocene and Early Eocene, horses had lost their reduced second and fourth toes and were left with only the third, middle toe to stand on, like modern horses do. Today, there is only one species of horse left over from this rich family tree: *Equus ferus*. The general trend in the evolution of the horse includes an increase in size, longer legs and head, a more complex brain, and wider incisor teeth and increased molarization, making them well adapted for a grazing lifestyle (Figure 9.37).

Rhinoceroses are a superfamily that include rhinos of today as well as a variety of extinct forms that all diverged from grazing perissodactyls sometime during the Oligocene (Figure 9.38). Early rhinoceroses were small and lacked horns. One extinct lineage of rhinoceroses was the *Hyracodontidae*, which started out rather small and spry. They later evolved into long-legged, hornless giants, such as the Asian, Eocene-Miocene *Indricotherium*, which reached heights of eighteen feet at the shoulder and a weight just short of twenty tons, making it the largest land mammal ever (Figure 9.39).

The artiodactyl (even-toed) ungulates began to outnumber and out-diversify perissodactyls midway through the Cenozoic. Some artiodactyl lineages have gone extinct, but many more are still with us today. Examples of extant artiodactyls include camels, pigs, bovids, giraffes, hippos, and deer. Camels evolved in the Late Eocene and are one of the oldest members of modern artiodactyls. One of the earliest camels is the *Protylopus*, which lived in the vicinity of South Dakota in the Eocene. *Protylopus* and its close relatives were only about the size of a rabbit. Its neck and legs were short compared to its body. By the Miocene, the direct ancestor to modern camels,

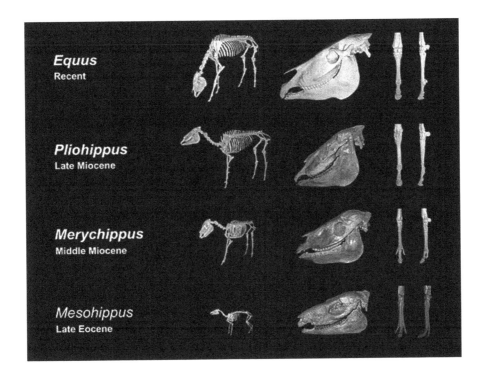

Figure 9.37: Examples of fossils showing the progressive evolution of the horse.

Procamelus, also evolved in the vicinity of South Dakota, Montana, and Wyoming. The way camels spread to the rest of the world was opposite of many migrations; they moved *westward* into Asia across the Bering Strait when the land bridge formed during the Pleistocene. They must have run into many animals migrating in the opposite direction. Camels also spread down into South America when the isthmus formed and evolved into llamas and alpacas. Today, camels show many adaptations for withstanding heat and lack of water. Contrary to popular belief, the humps of a camel are not used to store water; instead, they are concentrated fat reserves so that the rest of the body can stay lean and less insulated. Today's camels include the one-humped *Dromedary* from northern Africa and the Middle East and the two-humped *Bactrian* from Central Asia.

Bovids (family Bovidae) are artiodactyls that include impalas, antelopes, gazelles, goats, sheep, bison, cattle, deer, and wildebeest, to name a few. They evolved in the Eocene in Asia from a small, gazelle-like ancestor and spread to North America via the Bering land bridge. Bovids, as well as giraffes, are ruminants, which are artiodactyls with a digestive system specialized for extracting nutrients from grasses and leaves. Most species regurgitate the vegetation to chew on some more (as a cud) after it's been fermenting in their stomach

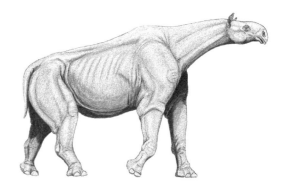

Figure 9.38: *Indricotherium (Paraceratherium)*, the largest land mammal ever.

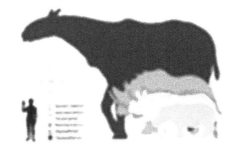

Figure 9.39: Size comparison of large mammals of the Rhinoceros family.

Figure 9.40: Skull of an *Oreodont* (*Hyposiops breviceps*).

for a while. Giraffes, which compose a different family than the Bovids, evolved in the Miocene from horned, deer-like ancestors in Africa called *Climacoceratids* (ladder-horned). The traditional view of giraffe evolution is that their long necks are an adaption to reach higher and higher tree limbs to eat the leaves. This is likely just one of the reasons why their necks got longer. A new theory that came out in the mid-1990s claims that giraffes with long necks are sexually selected through competition between males; the ones with large necks typically win the fight, which involves slamming their necks together.

Not all artiodactyls made it to the modern day. *Synthetoceras* were strange-looking deer-like ungulates from the Miocene with a long, forked horn protruding from just above their noses and two smaller horns on their heads. Oligocene four-toed *Oreodonts*, on the other hand, looked like a cross between a pig and a small horse, but with fangs (Figure 9.40). During the Oligocene and Miocene, they roamed the grasslands (and in some lineages, the marshes and swamps) of North America. Both *Synthetoceras* and *Oreodonts* are related more closely to camels than to pigs and horses. The omnivorous Eocene to Miocene *Entelodonts* were some of the fiercest-looking two-toed artiodactyls, and are informally referred to as "Hell Pigs" (Figure 9.41). They were among North America's apex predators in the Miocene and Oligocene, living in plains and forests and scavenging, hunting prey, and consuming roots and tubers as well. They were about seven feet long and had long jaws lined with sharp canines, incisors, and grinding teeth. They looked like a cross between a hyena and a horse and are among the clade from which **cetaceans** such as whales evolved from.

When grasslands spread during the Oligocene and Miocene, artiodactyls became more successful than the perissodactyls due to their complex digestive systems and ability to process the lower-nutrient grasses.

ELEPHANTS

Another familiar placental to evolve in the Late Eocene was the *Moeritherium*, one of the earliest members of the elephant order (Proboscidea). *Moeritherium* was small, about three feet tall, had a set of small, tusk-like teeth on the upper and lower jaws, and likely ate vegetation that lined the shores of lakes and rivers (Figure 9.42). Not long after, still in the Late Eocene, *Paleomastodon* appeared, possibly as an offshoot of the *Moerithium*, and by this time had begun to display longer tusks and a more elongate nose, although nothing yet like the trunks of their mastodon descendants, which lasted until only a few thousand years ago (Figure 9.43). True mastodons evolved in Africa during the Oligocene and spread across the Northern Hemisphere by the Miocene, in the process diverging into two truncated lineages ending in the American mastodon and the *Deinotheres*, which, during the Late Miocene, was

Figure 9.41: *Entelodont,* also known as the "Hell Pig."

Figure 9.42: *Moeritherium* hanging out by the water. These were some of the earliest members of the elephant order.

Figure 9.43: *Palaeomastodon* began to show the elongate trunks and tusks seen more prominently in their descendants.

one of the largest land mammals. Unlike mastodons, which had large tusks that curved upward and outward, Deinotheres had smaller tusks that curved downward. They went extinct at the start of the Pleistocene ice age, unlike the mastodons, which survived almost to the present day.

Mammoths came much later than the first mastodons, appearing in the Late Miocene as an evolutionary offshoot of *Gomphotherium*, a family of proboscidea that sported four tusks: two thick, blunt upper tusks that curved downward and a smaller set of closely spaced lower tusks that were spatulate in shape, sort of like shovels (Figure 9.44). Gomphotheres spread across

Figure 9.44: *Gomphotherium*, which gave rise to mammoths and modern elephants.

East Africa, Europe, and Asia, and finally migrated across the Bering land bridge to North America during the Pleistocene (Figure 9.45). Along the way, Gomphotheres diverged into mammoths by the Late Miocene and into modern elephants during the Pliocene. This, of course, means that elephants are more closely related to mammoths than to mastodons.

There are several ways to distinguish mammoths from mastodons. Mammoths were longer and more graceful-looking than mastodons, which tended to be shorter and stockier. Both were similar in size to modern elephants, and the size differences between elephants, mammoths, and mastodons are so little that scientists sometimes argue about which was bigger than which, although most claim that mammoths were taller. Mammoth molars are typically flat-topped with a folded, or corrugated, surface, like those of elephants. The molars of mastodons, however, have large cusps instead. Mammoths had a more dome-shaped forehead than mastodons, whose heads were a little more streamlined. Both had large tusks that pointed out and upward, but those of mammoths were a little longer and slightly more curved. Today, the only living proboscides are the African and Indian elephants, although one can still find relatively fresh frozen mammoths in the ice fields of high northern latitudes.

BATS

Bats (order Chiroptera) are special because they are the only mammals that can fly using their own bodies, an adaptation convergent with that of insects, birds, and Mesozoic reptiles. The earliest bat fossils are from the Eocene,

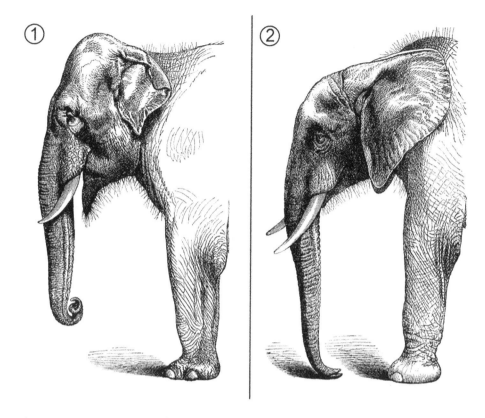

Figure 9.45: Comparison of the anatomy of the head and forepart of: 1) the Asian elephant *(Elephas maximus)*; and 2) the African elephant *(Loxodonta africana)*.

and their fossil record prior to that period is, so far, blank. This doesn't seem so surprising when you consider how delicate and difficult to fossilize the bones of bats are. Given that Eocene bats had by that time acquired highly specialized wings, the direct ancestors to bats must have spanned back perhaps into the Late Cretaceous.

One of the earliest bat fossils, *Onychonycteris finneyi*, came from 52.5-million-year-old Early Eocene rocks from Wyoming, and the small ear openings in its skull indicate that flight may have evolved *prior* to the use of echolocation, which is used to home in on insects at night, as seen in modern bats. These early bats had five clawed fingers; modern bats, on the other hand, have one or two digits on their hands. These claws indicate it was a good climber in addition to flier.

A contemporary of *Onychonycteris*, *Icaronycteris did* have the skull shape indicative of echolocation, so it was likely that by this time, some bats were already adopting a nocturnal lifestyle (Figure 9.46). The earliest bats were likely active during the day but may have been outcompeted for food, and probably eaten, by birds in the Early Cenozoic. This perhaps forced them to become nocturnal, which led to the development of echolocation to hunt at night.

The evolution of bats is poorly understood due to the lack of fossils. It is theorized that insectivorous bats branched from arboreal, shrew-like mammals during the Late Cretaceous or Paleocene. Interestingly, fruit bats (megabats, or flying foxes), which show up

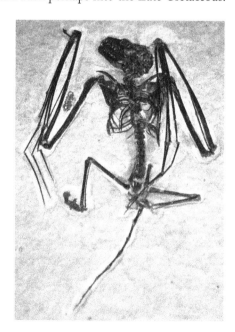

Figure 9.46: Fossil of *Icaronycteris*, an early bat.

Figure 9.47: Two main types of bats; Order Chiroptera (left) and the suborder Megachiroptera (right).

in the fossil record around thirty-five million years ago, are very different in many respects from insectivorous bats (Figure 9.47). Some scientists theorize they evolved from an early primate ancestor and subsequently underwent convergent evolution toward the insectivorous bats of a different ancestry. Fruit bats have a very poor fossil record; only three fossils exist. One is the Oligocene *Archaeopteroptus transiens*, and the others are from Miocene rocks in Africa and Poland.

CARNIVORES

The order Carnivora includes the pinnipeds (discussed in the Marine Mammals section), dogs, cats, bears, weasels, foxes, and raccoons. Members of this predatory order are quite diverse but unified in their dietary preference: meat. The earliest carnivorous mammals may have evolved from a group of Early Paleocene insectivores. By the Eocene, several lineages of carnivores had been established. One important, but now extinct, lineage is the ***Creodonts***.

The traditional viewpoint was that modern carnivores evolved from members of the *Creodonta* order, a group of carnivores that evolved in the Paleocene and went extinct by the Miocene (Figure 9.48). Creodonts are subdivided into two families. The *Oxyaenidae* lived in North America and Eurasia and were superficially cat-like both in appearance and in hunting style, using their spry, muscular bodies, sharp claws, and canines to ambush prey. The *Hyaenodontidae*, on the other hand, were more dog-like both in the shape of their face and their overall posture and gait. They were, on average,

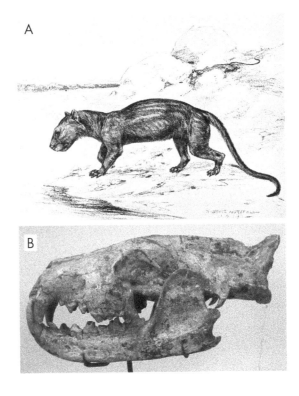

Figure 9.48: Two families of Creodonts: Illustration of *Oxyaenidae* (top) and a skull of *Hyaenodontidae* (bottom).

Figure 9.49: *Andrewsarchus, one of the last of the mesonychids, reached lengths of up to 13 feet long.*

about the size of dogs, but some species reached lengths of ten feet. Their dentition differed from true dogs in that they lacked molars and had a more restricted diet. *Hyaenodont* fossils are found in North America, Eurasia, and Africa.

Mesonychids are an extinct group of early (possibly the *earliest*) carnivores. They may fall into the direct ancestry of *archaeocetes* (and later, whales), although at this point in their evolution, they were still very much terrestrial. They had strong jaws capable of tearing meat and crushing bone, a wolf-like body, and a predatory lifestyle. However, rather than having claws, their toes were hoofed, indicating an ungulate ancestor not only for mesonychids but for cetaceans such as whales. A well-known mesonychid is *Andrewsarchus*, which represented one of the last, and largest, groups of mesonychids to evolve prior to their extinction (Figure 9.49). The fossil consists of just the head, but calculations based on its dimensions indicate it may have been over thirteen feet long and six feet high at the shoulder, with possibly a cat- or bear-like body.

Miacids, like creodonts and mesonychids, arose in the Paleocene and lasted until the Eocene. Today, it is generally thought that modern carnivores evolved from miacids during the Eocene, and not the creodonts. Miacids were generally small, and many species were arboreal. They had long, slender, weasel-like bodies and rather short skulls. It is thought their diets consisted of smaller vertebrates and insects. By the Oligocene, the direct lineages of cats, dogs, and mustelids (weasels, skunks, raccoons, etc.) had arrived. Not long after, bears and pinnipeds (seals, sea lions, and walruses) evolved.

MARINE MAMMALS

All amphibious or completely marine mammals are descended from terrestrial, four-limbed placental mammals. Not all radiated into the seas at the same time, however; it appears to have happened several times throughout the Cenozoic.

Cetaceans include whales, dolphins, and porpoises, all of which are highly intelligent marine mammals. The evolutionary history of dolphins and porpoises probably began from the same terrestrial roots as whales, which are focused on and discussed below. Some scientists suggest that whales evolved from Eocene *anthracotheres*, a semi-aquatic artiodactyl relative of today's hippos that fed on aquatic plants near the shores of rivers and lakes. According to other theories, whales evolved from a Paleocene/Eocene land-based quadruped, most likely small, deer-like animals such as *Indohyus* that may have fled to streams and lakes for protection against predators. *Pakicetus* is considered by many to be the first whale (Figure 9.50). *Pakicetus* was an Early Eocene wolf-sized carnivore that ate fish as well as terrestrial prey. Its head had the long, drawn-out shape and type of ear bones like those of today's whales. A descendant of *Pakicetus* fits loosely into the evolutionary trend of whales; the Eocene *Ambulocetus* was able to walk on land as well as swim easily in the water. It had the long, pointy head seen in the *Pakicetus* and had already adapted to being able to swallow underwater. Another possibility is that the ancestors to whales lie in the mesonychids, as discussed earlier.

The oldest fossils that are comfortably part of the whale lineage are the *Archaeocetes*, which at this point still retained many features left over from their terrestrial ancestors. Early Eocene *archaeocetes* had differentiated teeth, nostrils near the front of the mouth (as opposed to a blowhole), and hind limbs attached to the vertebrae by a

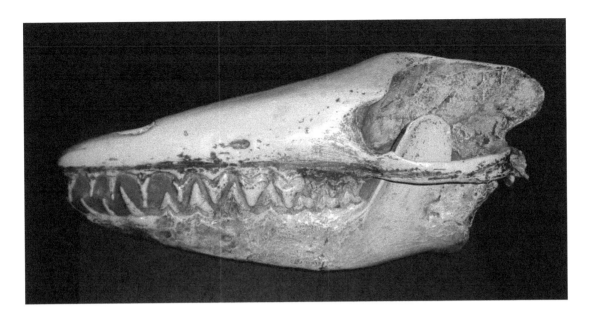

Figure 9.50: Skull of *Pakicetus*, the first true whale. Note the elongate skull and high nostrils, adaptations seen more pronounced in modern cetaceans.

sacral joint. Early *archaeocetes* gave birth on land, where they spent much of their time; they likely went into the water to feed.

By the Late Eocene, whales had taken on more of a modern form and purely marine lifestyle, such as the *Basilosaurus. Basilosaurs* had relatively small front flippers and even smaller vestigial hind legs. Its head and snout were small and tapered, but it had a long, slender body, up to sixty feet in length. It had sharp, pointy teeth and likely fed on fish.

Modern whales still retain vestigial pelvises and femurs; they are completely detached from the spine and serve no apparent function at all. Some whales (*Odontocetes*) still retain blunt, rounded, peg-like teeth, while others, such as baleen whales, have no teeth and filter-feed plankton using baleen plates.

Pinnipeds ("fin-feet") are carnivores that include today's seals, sea lions, and walruses. The fossil record of pinnipeds is somewhat poor; it is speculated they evolved sometime in the Eocene or Oligocene from an ancestral stock unrelated to those that sprang forth the cetaceans. Some researchers suggest these ancestors were more related to bears, while others indicate they were from the Mustelidae family. Primitive Miocene species like the Canadian *Puijila darwini* were most likely some of the earliest *transitional* seals, as they were seal-like in size and overall body

Figure 9.51 The Early Miocene *Puijila darwini*, an early transitional seal.

shape but still retained limbs better suited for land than for water (Figure 9.51). Otters, which are not pinnipeds, but rather part of the Mustelidae family, are thought to have arisen in the Miocene from fish-eating amphibious mustelid ancestors.

EFFECTS OF THE PLEISTOCENE ICE AGE[1]

THE PLEISTOCENE ICE AGE

The Pleistocene began about 1.9 million B.P. (1.9 million years Before Present) and ended about 10,000 B.P. with the world wide extinction of many large mammals. The time since the end of the Pleistocene is termed the Holocene in North America. The term Holocene is generally equivalent to the term Recent.

Portions of time that include glacial advances are called glacial stages and long-term temperate periods are called interglacial stages (Figure 9.52). These stages have names that differ in different parts of the world (Figure 9.53). For instance, the Wisconsinan, which is the last glacial stage of the Pleistocene in North America, is called the Weichselian in northern Europe, the Devensian in the British Isles, the Tali in China, and the Gamblian in East Africa.

With the advent of radiocarbon dating in the 1950s, it became possible to obtain dates for the terrestrial Pleistocene based on the analysis of fossil material. Correlation of terrestrial and marine deposit was not possible

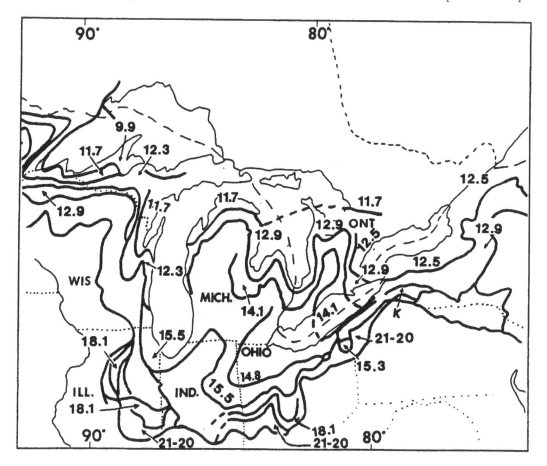

Figure 9.52: Positions of the Pleistocene ice margin in the Great Lakes region in thousands of years B.P.

1   J. Alan Holman, "The Pleistocene Ice Age," *In Quest of Great Lakes Ice Age Vertebrates,* pp. 10-14. Copyright © 2001 by Michigan State University Press. Reprinted with permission.

until sedimentary cores from the sea bottom were collected that represented continuous deposition over long periods of time. These cores showed that alternating biotas of warm-water and cold-water organisms, mainly foraminifera, existed. Coupled with studies of oxygen isotopes and magnetic reversals, these cores provided an accurate time scale for the expansion and contraction of continental ice sheets.

HOLOCENE	Bronze Age and Later Industries
	Neolithic (New Stone Age)
	Mesolithic (Middle Stone Age)
PLEISTOCENE	Palaeolithic (Old Stone Age)

Figure 9.53: Cultural industries in Europe.

At present, the sequence of cold and warm cycles shown by evidence from the sea bottom provides the best information about what was probably happening on land. Good correlations between marine and terrestrial Pleistocene events have been established for the later parts of the Pleistocene throughout the world.

It has been recently demonstrated that many glacial advances and withdrawals (Figure 9.52) as well as climatic fluctuations occurred within the classic Pleistocene stages shown in Figure 9.53. Nevertheless, these classic terms are still used internationally to depict generally cold (glacial) and temperate (interglacial) stages.

An explanation for the cause of glacial ages has been searched for ever since the learned Louis Agassiz expounded in the middle of the nineteenth century on the "Great Ice Age." Nevertheless, a widely accepted theory (such as natural selection to explain evolution or plate tectonics to explain continental drift) has not emerged to address Pleistocene or earlier glacial events. Such a theory could involve cyclic activities in the atmosphere due to changes in solar output, irregularities in Earth's orbit and rotation, volcanic periodicity, or even variations in Earth's magnetic field.

In Europe, special terms are used for human cultural industries of the Quaternary. These names are mainly based on human artifacts. Hand axes and spear points were common in the Pleistocene, whereas more sophisticated artifacts occurred in the Holocene. Palaeolithic (Old Stone Age) industries occurred in the Pleistocene, whereas Mesolithic (Middle Stone Age), Neolithic (New Stone Age) and Bronze Age and subsequent cultures occurred in the Holocene (Figure 9.55). Since humans did not arrive in North America until the late part of the Late Pleistocene, the European Paleolithic subdivisions do not apply in the Americas.

Ice sheets of the past were originally recognized on the basis of evidence from ice formations of the present. Today, these ice formations are restricted to high latitudes and high altitudes. Giant continental ice sheets still cover Greenland and Antarctica, and the Arctic Ocean is frozen over permanently. Moreover, extensive glaciers occur in the southern Andes, the Rockies, the Alps, and the Himalayas.

Both modern and ancient ice sheets leave undeniable evidence of their movements. This evidence includes scoured and polished bedrock, U-shaped valleys formed by erosion, and relocated foreign rocks called erratics that were plucked from their original location by the moving ice. Beyond the extent of the glaciers, one finds sands and gravels from glacial meltwaters as well as landscape features.

Studies of this kind of evidence have established a detailed picture of where the ice was during the Late Wisconsinan as well as equivalent ages such as the Late Devensian and Weichselian. All of the modern ice sheets and glaciers expanded tremendously during this time. Giant ice sheets covered the northern part of Europe, including most of the British Isles and all of Finland and Scandinavia. In North America, the ice extended as far south as southern Illinois, Indiana, and Ohio (Figure 9.54). Oddly, Alaska and part of the Yukon and Northwest Territories remained mainly unglaciated. In the Southern Hemisphere the ice cover increased in the Andes, there was a significant increase in the amount of the sea ice in Antarctica, the glaciers in Africa extended lower, and even Tasmania, which has no ice cover today, had an ice cap.

Figure 9.54: Maximum extent of the Illinoian and Wisconsinan glaciation in the Great Lakes region. This is the farthest south the ice penetrated in North America in the Pleistocene.

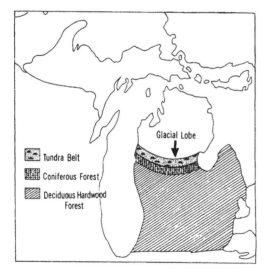

Figure 9.55: "Stripe hypothesis" illustrated. Biological communities in an orderly advance ahead of the ice sheet during the Pleistocene in Michigan.

THE PLEISTOCENE IN NORTH AMERICA

Two giant ice masses existed in the North American Pleistocene. The largest of those was the Laurentide Ice Sheet, which extended from Nova Scotia and the northeastern United States across the continent to western Canada. The Laurentide Ice Sheet impacted the Great Lakes region in the Illinoian and Wisconsinan stages of the Pleistocene. A smaller mass, the Cordilleran Ice Sheet, covered the mountain ranges of the Northwest from Montana and Washington up to the Aleutian Islands.

In North America, the ice sheet penetrated farthest in the central Great Lakes region. The most southern penetration took place during the Illinoian glacial stage when the Laurentide Ice Sheet extended to southern Illinois, Indiana, and Ohio (see Figure 9.54). During the Wisconsinan, the maximum penetration of the ice did not extend quite as far south in these states (see Figure 9.54).

EFFECTS OF THE ICE SHEET

The general effects of the Laurentide and Cordilleran Ice Sheets in North America are summarized below.

COMMUNITY DESTRUCTION

When ice sheets advance, vast areas of habitat are destroyed. Some simple organisms, plants, and insects can live on or within the ice, but major life forms are obliterated from areas lying under the masses of ice. For in stance, in the central Great Lakes region, in the Late Wisconsinan, about 20,000 B.P., the ice sheet blanketed huge areas of habitat and changed the nature of habitats in the southern, unglaciated portions of Illinois, Indiana, and Ohio.

Additional advances occurred at intervals of about 1,000 years before a final advance extended to northern Indiana and Ohio about 15,000 B.P. About 14,800 B.P. the ice started its final withdrawal in the Midwest and exposed the land for recolonization by animals and plants.

TOPOGRAPHIC CHANGES

The Pleistocene ice sheets drastically changed the landscapes over which they passed. The thickness of the North American ice sheets varied from place to place. It has been estimated the average thickness was about 1.25 miles and that in places it was 2 miles thick or more. Valleys, ridges, various types of small hills, lakes (including the Great Lakes), streams, swamps, and bogs were all produced by ice sheet activity. All of these features influenced the ecology of Pleistocene vertebrates.

CLIMATIC CHANGES

Certainly the movements of the ice sheets produced dramatic climatic changes. Nevertheless, the classic idea of alternating cold glacial and warm interglacial climates is now considered to be oversimplified. Modern evidence in North America indicates that climates were cold in areas near the ice sheet borders, but in the central and southern United States the climate is believed actually to have been more equable than it is today, with warmer winters and cooler summers. This theory is termed the Pleistocene climatic equability model.

VEGETATIONAL CHANGES

The advancing and retreating ice sheets of the Pleistocene altered vegetational communities. The classic idea was that major vegetational associations were caused to with draw southward in bandlike units by the advancing ice and that these units were caused to move northward in the same way by ice sheet retreats (Figure 9.55). It was thought that in the eastern United States during glacial times, a barren tundra association existed in a deep band south of the glacial front and that this was followed by a deep band of coniferous forest that graded into temperate deciduous forest, which penetrated far into the Southeast. This classic concept is often referred to as the stripe hypothesis.

The modern theory, however, is that during a large portion of the Pleistocene, a cold climate and tundra or coniferous vegetation existed in areas rather near the edge of the glacier but that the vegetational communities in the central and southern United States existed as a mixture of the original plants of the area coexisting with invading northern forms. This theory has arisen from the idea that plant and animal species reacted individually rather than as groups to Ice Age changes and that the mixed communities in the region would have been able to coexist in the equable climates of the time. The modern theory is often referred to as the plaid hypothesis.

SEA AND LAKE LEVEL CHANGES

During Pleistocene glacial times, so much of the world's water was bound up in the ice sheets that sea levels fell. For example, in Florida, the peninsula enlarged greatly as the sea withdrew during glacial times. On the other hand, during interglacial times, sea and lake levels rose considerably. In the Great Lakes region, the Great Lakes themselves rose and fell, as did large rivers such as the St. Lawrence.

REORGANIZATION OF BIOTIC COMMUNITIES

When the great ice sheets melted, they left a mass of virtually sterile mud, silt, sand, and gravel in their wake. This new material had to be recolonized by pioneer species during the period of ecological succession that must have followed. Certainly there must have been a considerable lag time between the retreat of the Pleistocene ice sheets and the development of fully developed and stable plant communities.

10

Figure 10.1: Skull of *Australopithecus sediba*, a recently-discovered possible link to the genus *Homo*.

ORIGINS OF OUR SPECIES

AN ASTEROID CRASH ALLOWED MAMMALS TO DIVERSIFY[1]

Humans have a long evolutionary history. Species closely related to humans began to emerge 5 to 7 million years ago, and our ape ancestry goes back 20 million years or more. The root of the branch of the mammalian tree that led to humans goes back at least 65 million years, and of course mammals emerged long before that. It is easy to be confused by details when studying human evolution within this vast stretch of geologic time. The cast of fossil characters is large and remote in time. Their names are unfamiliar and their traits are diverse and complex. Human evolution is an active field of research and new findings sometimes complicate the picture instead of clarifying it. DNA studies answer some questions about the relationships of fossil humans to *Homo sapiens*, but they raise others. Even the definition of the word "human" often is revised by fresh observations of the behaviors of our closest relatives. This chapter will help you to understand many of these issues and complexities. One starting point is that humans are mammals.

The earliest mammals evolved from reptile ancestors more than 120 million years ago. The earliest **primates,** however, did not evolve until around 65 million years ago. What environmental conditions favored the evolution of our primate ancestors? Many researchers believe that the extinction of dinosaurs allowed the evolution of new kinds of mammals, including primates. So, let's begin this discussion of human evolution by considering one wide scale phenomenon that led from early mammals to you, *Homo sapiens*, reading this page at this moment.

If you could travel 68 million years into the past, Earth and its inhabitants would be far different than they are today. At that time temperatures were uniformly tropical—even in what is today Antarctica. The positions of the land masses were quite different, too. Your attention might be caught by a tropical landscape with subtly different plants (no flowers, for instance) and dramatically different animals (no monkeys, lions, or elephants).

---

1 Jan Jenner and Joelle Presson, "The Evolution of Homo sapiens," *Biology: The Tapestry of Life*, pp. 422-425. Copyright © 2014 by Cognella, Inc. Reprinted with permission.

Sixty-eight million years ago large reptiles of all sorts ran in herds and family groups, alert for attack by efficient reptilian predators. Gargantuan herbivores lifted their heads to the tops of *Sequoia* trees to graze on tender, new needles. Some dinosaurs excavated nests and laid their eggs in them. Some may even have helped their young to hatch, in the way that crocodiles and alligators do today. The world 68 million years ago was *Jurassic Park* come to life. Small mammals, some the direct ancestors of modern primates, clung to life around the margins of this world dominated by dinosaurs and other reptiles. Then something happened that gave our small primate ancestors their chance to diversify and evolve.

Around 65 million years ago a mass extinction occurred, often called the **K/T event**. This mass extinction dramatically changed Earth's cast of characters by wiping out most land animals that were larger than a rat. The K/T event eliminated giant reptiles and allowed smaller forms of life, like mammals and birds, to eventually assume the roles of major players. ...

Not all life was extinguished in this devastating event. Some plants that died left spores or seeds that germinated when the environment returned to better conditions. Scavengers and insectivores that could subsist on the fringes of the changed world became the nucleus of new kinds of animals. With the dominant dinosaurs and other large reptiles wiped out, small-sized primates and other families of mammals that survived could diversify. So, the K/T event blasted open a new evolutionary door. Mouse-sized primates crouched behind it and they began to take advantage of these newly available ecological opportunities. Many new species of primates emerged in the millions of years after the K/T boundary.

The K/T event is an example of how an unexpected and rare, but cataclysmic, event can result in dramatic changes in the evolution of life. The K/T event was caused by the impact of an asteroid. Other unexpected events are volcanic explosions, earthquakes, or tsunamis. Unexpected events that influence the whole of life on Earth are rare, but such events often can be found in more local venues throughout the history of life on Earth, and they can dramatically change the pace and direction of evolution.

GEOLOGIC AND CLIMATE CHANGES INFLUENCED PRIMATE DIVERSIFICATION[2]

It is worth learning something about the structure of continents and how they move. As you may know, Earth is a spherical, semi-liquid layer cake in which less dense, cooler, and more rigid continents ride upon underlying *tectonic plates* (Figure 10.2). These plates, in turn, slide slowly and in different directions upon a deeper, denser, hotter layer of semi-liquid rock. Movements of the continental plates are propelled by two processes: sea floor spreading and subduction. In *sea floor spreading* new rock seeps up out of Earth's liquid mantle, creating currents in the semi-liquid rock beneath continental plates. *Subduction* happens when one continental plate collides with another and moves below it. Alternately, one plate can rise up over another. Propelled by tectonic movements, Earth's continental plates have been drifting, colliding, merging, and separating for as long as geologists can estimate. Continental plates are moving right now and geologists expect that continental drift will continue indefinitely.

What do movements of continental plates have to do with primate evolution? Think again of dominoes falling as you read the following broad-brush explanation:

- Tectonic activity changes the relative positions of continental masses and this, in turn, affects global climate.

---

2   Jan Jenner and Joelle Presson, "The Evolution of Homo sapiens," *Biology: The Tapestry of Life*, pp. 425-427. Copyright © 2014 by Cognella, Inc. Reprinted with permission.

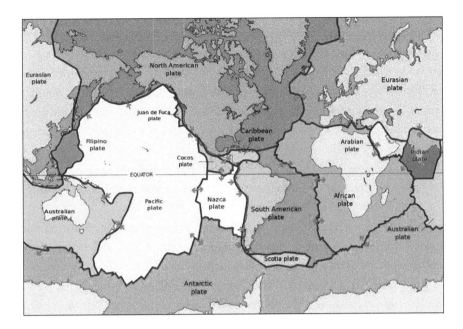

Figure 10.2: Tectonic plates. Like an enormous jigsaw puzzle, Earth's tectonic plates lie beneath Earth's land masses and oceans. The arrows indicate direction of plate movement.

- Changed climate affects the geographic distributions of major plant biomes. Tropical forests, the ancestral homes of species of apes, advance and retreat as do drier grassland biomes.

- The changed vegetation favors some kinds of animals while it restricts or eliminates others.

Now let's add some specific details to the broad-brush picture of these geologic and climatic changes that have influenced primate—including human—evolution. About 54 million years ago the Indian continental plate collided with the Asian continental plate. This massive collision eventually caused the rock layers of the Himalayas to crumple and rise 5 kilometers (a little over 3 miles). The uplift of the Himalayas continued until about 15 million years ago. Not only were the peaks of the Himalayas uplifted, but the Himalayan plateau also rose to an elevation of about 5,000 m (16,400'). This is no small geographic feature: the Himalayan plateau is about equal in area to the area of the lower 48 U.S. states. And, as you might expect, the rise of such a huge land feature altered the pattern of global air currents and ultimately climates worldwide.

CLIMATE CHANGES ALTER PLANT DISTRIBUTIONS

Before the collision of the Indian and Asian plates, Earth was experiencing 100,000 years of extreme global warming. Sea surface temperatures had risen over 8°C (46.4°F) in a little over 10,000 years. The rise of the Himalayan plateau changed this. It channeled cold Arctic air toward lower latitudes and global temperatures began to drop. In addition, rainfall patterns were disrupted. Heavy rains on the Tibetan plateau and in the nearby Himalayas weathered the rocks and exposed them to the atmosphere. Chemical reactions in rocks and soil removed carbon dioxide from the atmosphere, intensifying the drop in global temperatures. By about 36 million years ago an ice sheet covered Antarctica and in North America the average annual temperature had dropped by 12°C (53.6°F). A series of glacial periods of varying intensity followed. The last of these ended about 23 million years ago and world climate subsequently warmed. Of course, these fluctuations in temperature affected vegetation worldwide and changes in distribution of plants had wide ranging effects upon animals.

Plant distribution is in large part governed by climate. For instance, as Earth warmed 55 million years ago, tropical evergreen forests expanded. Eventually they covered even high latitudes from Southeast Asia, Europe and North America. Then about 36 million years ago, as the planet cooled dramatically, tropical forests were restricted to low latitudes. Over the next 13 million years warm and cold periods alternated and tropical forest cover continued this pattern of expansion and shrinking. Twenty-three million years ago, the climate had warmed and tropical and subtropical forests expanded in Africa and Eurasia, reaching up into Siberia. This set the stage for diversification of primate species that led to the ancestors of apes, and eventually to the ancestors of apes and humans.

Geological and climate changes also are associated with more recent events in human evolution. The Rift Valley in Eastern Africa, one of the probable areas where early humans first evolved, underwent corresponding climate changes—periods of tropical warmth and humidity alternating with drier periods when tropical forests retreated. There also were smaller scale African geological changes around a million years ago as the Nubian and Somalian tectonic plates underlying west and eastern Ethiopia pulled apart, creating the Rift Valley. Overall, Africa was colder and drier than in earlier times, but the climate was also more variable, with dry periods interspersed with wetter periods. As the climate cooled and dried, the African tropical forests—where early primates and apes evolved and thrived—gradually gave way to dry woodlands and grass-covered savannahs. As the environment shifted from forest to grassland and savannah, early human-like species continued to evolve adaptations, such as tall stature and loss of body hair to facilitate heat loss. Some scientists speculate that the drier but more variable environments 2 to 5 million years ago allowed species with unique physical, intellectual and cultural traits to thrive. Individuals and populations that could manipulate their environments, learn new skills, and teach skills to their children had a distinct advantage in a challenging environment. The ability to stand fully upright, run efficiently, and use group strategies to hunt big game would have resulted in an increase in reproductive success. Of course, it is difficult to rigorously test such hypotheses, but there is little doubt that geological change, climate change, and their environmental consequences played a large role in early human evolution. ...

EARLY PRIMATES

The earliest primate-like mammals in the Paleocene were very different from the apes that evolved later in the Cenozoic. We use the term "primate-like" because there still exists a great amount of ambiguity about their taxonomic placement. For instance, there are two major contenders for what some scientists call the first true primates. As postulated by several scientists based on its primate-like teeth, the Early Paleocene *Purgatorius* (Figure 10.3) may be the first primate.

It lived in trees and probably ate insects and fruit, the swollen ovaries of the relatively new branch of plants called angiosperms. This small, agile, shrew-like placental was one of the many species to co-evolve into a symbiotic relationship with angiosperms. Now that trees were bearing fruit, there was a reason to hang around up in the branches; the trees, as a result, had a method by which fruit-borne seeds could be dispersed by the organisms that ate them. The difficulty of plucking fruit from an often thin, frail branch high up in the air likely caused descendants of *Purgatorius* to respond with adaptations such as stereoscopic vision, prehensile tails, and opposable thumbs, the latter being a trait utilized by modern humans for countless purposes.

The slightly younger, approximately 55-million-year-old insectivorous *Archicebus* from Asia was more similar to modern tarsiers than the rodent-like *Purgatorius* and is instead considered by other scientists to be the first true primate. A contemporary of *Arhicebus* was the North American (and later European) *Plesiadapis*, which was larger than the *Archicebus*, reaching lengths of about two feet. Also arboreal, *Plesiadapis* differed from earlier "primates" in that it had teeth more suited for an omnivorous diet. Recently, it has become the poster child for early primate-like mammals; recent studies often place all of these Paleocene arboreal mammals in a group called Plesiadapiformes,

Figure 10.3: Illustration of *Purgatorius*, one the earliest primates.

and not the Primate order. They had not quite evolved the stereoscopic vision and opposable thumbs of true primates, and their dentition, although geared towards a more omnivorous diet, still superficially resembled that of a rodent. Today's primates constitute an order that is subdivided into two suborders: the Prosimians ("before apes") and the Anthropoids ("higher apes"). Prosimians today, which include lemurs, lorises, and tree shrews, are small, arboreal primates and are most like the Plesiadapiforms in their form and lifestyle. Modern prosimians are more evolved, however, possessing opposable thumbs and stereoscopic vision. In the Eocene, prosimians occupied much of North America and Eurasia, but cooling during the Oligocene forced their tropical habitats to withdraw toward the equator. Today, they are mostly found in Africa, the East Indies, and the warm, southern parts of Asia.

The suborder Anthropoidea (monkeys, apes, and humans) evolved from the prosimians sometime during the Early to mid-Cenozoic. Molecular data leads some workers to contend that anthropoids may have evolved as far back as the Late Paleocene. Tarsiers, which evolved around that time, appear to have within their genetic codes some evidence of anthropoid DNA, suggesting the presence of anthropoids much earlier than other workers have considered. Some studies suggest that *Archicebus* was the branching point of both the prosimian tarsiers and the anthropoids. This notion is confounded by the recent ambiguity of exactly where tarsiers fit into the taxonomic framework. The evolution of early primates, and especially that of the anthropoids from prosimians, is still a hot issue in paleontology, and new studies will likely continue to yield contradictory results. Anthropoids today are split into two geographical groups, the "**Old World** Monkeys" and the "**New World** Monkeys." Of the two, the New World Monkeys, which include spider monkeys, howler monkeys, and marmosets, are more derivative of the earlier primates; they are generally smaller, with a flattish face, prehensile tail, and widely separated nostrils. Although evolving in Africa, this group had migrated into South America prior to the split-up of southern Pangaea and remains there today.

The Old World Monkeys, as their name implies, are endemic to the warmer regions of Africa and Asia and include, among others, baboons and macaques. They tend to lack prehensile tails and have closer-set, downward-pointing nostrils and the ability to convey a wider range of facial expressions. Anthropoids can also be split into three superfamilies: Cercopithecoidea (the Old World Monkeys), Ceboidea (the New World Monkeys), and Hominoidea (the "apes"). Hominoids consist of three families: Pongidae (the Great Apes), Hylobatidae (the Lesser Apes), and Hominidae. Humans are the only Hominids living today, but the group also includes our recent ancestors and their cousins, upon which the rest of the chapter will focus.

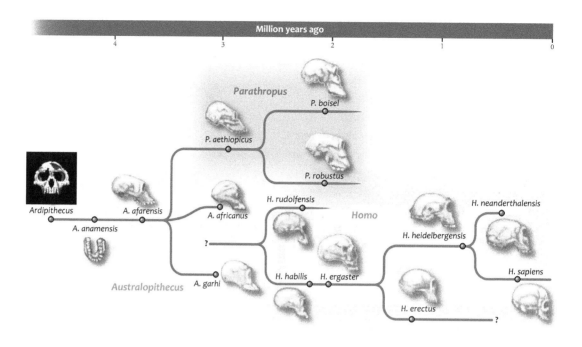

Figure 10.4: Human evolutionary history. Here is one hypothetical view of human evolutionary relationships. Note how human species once were quite numerous, with four or five species coexisting about 2 million years ago.

MOLECULES REFLECT EVOLUTIONARY HISTORY[3]

Until recently fossil evidence and anatomical similarities were the only lines of evidence that paleoanthropologists could use when trying to trace the course of human evolution. Now, studies of biological molecules and behavior can tell us about relationships between species of primates. Studies of blood proteins and DNA show that humans are one of several species of great apes. One of the astounding findings of DNA studies is that the DNA of humans and chimpanzees is 99% identical. Also, studies show that chimpanzees and gorillas make and use tools. When taught by humans in a laboratory setting, chimpanzees and gorillas can learn language—such as sign language. Some chimps have even been known to pass on sign language to their offspring. The phylogenetic tree shown in Figure 10.4 reflects the insights from anatomical, behavioral, and DNA studies. These studies have resolved that chimpanzees are more closely related to humans than either is to other species of apes. Here is a highly significant-but-subtle point: note that humans *are not* descended from chimpanzees. Instead, living humans share a common ancestor with living chimpanzees. Chimps and humans diverged from that common ancestor about 5 to 7 million years ago.

3   Jan Jenner and Joelle Presson, "The Evolution of Homo sapiens," *Biology: The Tapestry of Life*, p. 432-433. Copyright © 2014 by Cognella, Inc. Reprinted with permission.

HUMAN-LIKE SPECIES EMERGED GRADUALLY OVER THE PAST FIVE MILLION YEARS[4]

What was the common ancestor of humans and chimpanzees like? Did it resemble a modern chimp or a modern human or neither? Some people predict that the common ancestor was much like a chimp, on the assumption that modern chimps have not evolved since that common ancestor. The fossil record challenges these assumptions. Figure 11.3 shows an overview of what is currently known about human evolution, back to about four million years ago. To fully grasp this history, you will have to get used to some more new terminology. All apes including humans are phylogenetically related in a group called **hominids**. Some hominid species are so closely related to modern humans that they are included in the genus *Homo*, along with the human species, *Homo sapiens*. Other early hominids have more ape-like features, but still share more features with humans than they do with chimps. These are given various genus names but are grouped along with the genus *Homo* in a larger phylogenetic category called **hominins**. This taxonomic terminology can be confusing, so here is a quick summary of some important terms:

- *Ardipithecus:* the earliest known hominin; this genus originated around 4 million years ago.
- *Australopithecus:* a genus of hominins that lived from 4.2 to 1 million years ago.
- *Paranthropus:* a hominin genus that diverged from *Australopithecus* and then died out.
- *Homo:* a large genus of hominins that emerged around two million years ago. *Homo sapiens* is the only surviving species of this genus.

One of the first things to notice in Figure 10.3 is how many different species have emerged along human evolutionary history. There is not a single "line" of evolution that leads from the common ancestor of chimps and humans to modern humans, as many people assume. Instead, the human evolutionary tree is more like a bush, with some twigs forming dead ends. Evolutionary biologists do not find this surprising. Like the evolution of most species, many hominin species have emerged, become extinct, and new ones arisen in a succession and proliferation of species. Today only one hominin species survives—us—but this was not always the case. Over the past 5 million years many human-like species lived in Africa, Asia, and Europe. The transition to modern *Homo sapiens* was not abrupt or distinct. Early hominin fossils have traits that resemble both humans and chimpanzees, and the emergence of fully modern human traits was a gradual process that extended over millions of years. Within this complex evolutionary history it is not yet possible to identify the exact path of human evolution that led from the common ancestor of chimpanzees and humans to modern humans. That does not mean we have no direct ancestors. Rather, it is difficult to identify our specific ancestors from fossil evidence. In the future new fossil finds will certainly help to resolve this uncertainty. Nevertheless, some clear trends in hominin evolution are evident.

As you read above, the diagnostic feature that distinguishes the hominins from hominids is the ability to walk upright. This trait evolved early in hominin evolution. To determine how a fossil individual moved, researchers study overall posture, hip structure, and the position of the foramen magnum—the large hole in the skull where the spinal cord exits and enters the vertebral column. Subtle details of angles of bones around foramen magnum can provide important clues to posture, as can structure of limb bones and foot bones. As you will see below, the earliest hominin fossils show evidence of upright posture and walking.

The size of the skull is one of the most important features used to track hominin evolution, because it reflects the size of the brain. Modern humans have relatively large brains—about 1,200 cm^3, while the average brain of a

---

chimpanzee has a volume of 400 cm³. So a modern human's brain is about three times the size of a chimp's brain. Large brains emerged gradually over hominin evolution, much more gradually than did upright posture. The earliest hominins had brains no larger than a chimpanzee's; some later hominins had brains even larger than modern humans' brains. This increase in brain size has been accomplished by an increase in the number of neurons in the brain. This may seem an obvious point, but more neurons mean more computational power. Humans really can think more deeply and understand more deeply than other primates because they have more mental computing power. But, the human brain is not just an enlarged version of a monkey brain. Nor is a monkey brain just a larger model of a rat's brain. The organization of the brain has changed over evolutionary history. For example, primates in general have relatively more brain area devoted to the higher perceptual and cognitive functions, especially those related to vision. The *Homo* species show the emergence of an asymmetry in brain function. In *Homo sapiens* the right side of the brain does not carry out exactly the same functions as does the left side of the brain. Our species has an elaboration of the brain areas that allow language, of course, but we also have enlarged brain areas involved in the recognition of faces. While it is important that your brain is larger than a chimpanzee's brain, it is also significant that your brain is organized somewhat differently than a chimp brain.

Modern humans have other anatomical features that emerged over the past two million years, and that are used to distinguish hominid and hominin fossils. Overall body height, the size and shape of teeth, the relative lengths of limbs, the position of the skull on the spinal column, and degree of flatness of the face all are important anatomical features used to classify the fossils that are clearly similar to modern humans, but not fully human. Table 10.1

Table 10.1: Important anatomical features used to classify fossils of hominids and hominins.

	CHIMPANZEE	MODERN HUMAN
Height	5´6˝ tall (males) Males are taller than females	5´10˝ tall (U.S. males) 5´4 ½˝tall (U.S. females)
Details of teeth and skull	Tooth row is U-shaped Canines self-honing Males have larger canines than do females Teeth and jaws larger No chin Large brow ridges	Tooth row is C-shaped Canines not self-honing No sexual differences in size of canines Teeth and jaws smaller Chin is present Brow ridges quite small
Length of limbs, details of digits	Longer arms, shorter legs Opposable thumbs and large toes	Shorter arms, longer legs Only opposable thumbs
Position of skull on spinal column	Spinal column exits at extreme rear of skull	Spinal column exits beneath skull
Shape of pelvis	Flatter, more blade-like ilia (pelvic bones); less evidence of weight-bearing	Ilia are bowl-shaped to bear weight
Flatness of face	Prognathous (forward-jutting) jaw	No prognathous jaw, flatter face

shows some of these features, and compares them to the traits of modern humans and modern chimpanzees. You will explore these more in the sections below.

Behavior is an important feature used to categorize and understand modern living species. In many ways human behavior is similar to behavior of other primates, and especially to behavior of other hominids. But in some ways human behavior is different. First, while chimps and other apes do use tools, *Homo sapiens* consistently uses sophisticated manufactured tools. Second, while chimps and gorillas can learn to use language, they do not do so routinely in the wild, and when non-human apes do learn language they never reach the sophistication of modern humans. These patterns are not surprising. You would not expect new traits to emerge in a species with no prior evolutionary foundation. Human tool use and language have built upon the traits of non-human primates. The key question is when did the modern human levels of these traits emerge? Did they emerge early as did upright walking, or did they emerge gradually as did a large brain? Even though there are no fossils of "behavior," the fossil record can address these questions.

One question that fossil evidence and molecular data can answer is when and where modern *Homo sapiens* emerged. Africa is the ancestral home of all hominin species. Many of the hominin fossils have been found in Africa. The common human-chimp ancestor lived in Africa, and the diverse human-like hominin species evolved in Africa. Our direct ancestors emerged in Africa, probably northern Africa, about 200,000 years ago. But, not all hominins stayed in Africa. While most of our ancestors stayed in Africa—lived their lives, and developed their cultures there—some ancestors moved out of Africa to colonize nearly all corners of the world.

EVOLUTIONARY HISTORY REVEALS MANY KINDS OF HUMANS

In your family album you may have a photograph of a long-lost great-great-great grandparent who came to America from the Old Country, seeking a better life. If you're lucky, you may have an oil portrait or a miniature, a tintype or a locket that shows the face of this ancestor, whom you've never met. Perhaps you can trace a family resemblance—a strong chin, ears like jug handles, a high forehead, perhaps unusually shaped eyebrows, or a characteristic nose. The next section of this chapter is a kind of human family album. But instead of pictures or photos of your relatives, you'll be considering fossilized remains of hominin relatives.

Ardipithecus. The ancient history of our species has long been mysterious and unknown, but in 2009 scientists described hominin fossils dating back 4.8 to 4.4 million years, a time not long after hominins split from the common chimp ancestor. *Ardipithecus ramidus* fossils were discovered in Ethiopia in the early 1990s, but the research team studied them for about 15 years before publishing their findings. The name *Ardipithecus* means "ape on the ground floor," indicating that this species is or is extremely close to the first hominid.

The first important conclusion about Ardi, as these fossils are called, is that they are not chimpanzees, but nor are they fully human. Ardi was roughly the size of a modern chimp, about 130 cm. tall (about 4') and weighed 50 kilos (about 110 pounds). Its brain was small, similar to that of a chimpanzee. Surprisingly, however, Ardi had a more upright posture than chimps, and skeletal features that were more human than chimpanzee. Although Ardi walked upright, she also moved on all fours in the trees, where she spent much of her time. Ardi's skull sat more upright on the spine than a chimp's does. Her face was relatively flat, and her canines were shorter and less sharp than the enlarged canines of a chimpanzee (Figure 10.5). Ardi probably was an omnivore, eating both meat and plants. These traits are more like a human than a chimpanzee. Were these traits of the common chimp/human ancestor, or did these traits evolve in early hominins? At this point scientists do not know. *Ardipithecus* fossils support the conclusion that both humans and chimps have been evolving for 5 to 7 million years and that their common ancestor was neither fully chimp-like nor fully human. Two species of *Ardipithecus* have now been found:

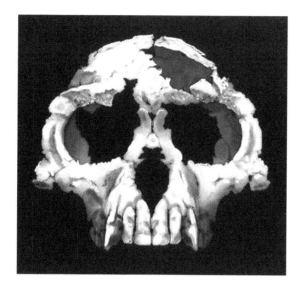

Figure 10.5: *Ardipithecus ramidus* skull. From this badly fragmented skull that has carefully been pieced together one can note that the teeth, especially the canines, are small, and so is the brow ridge. This may indicate that the skull belonged to a female. *Ardipithecus ramidus* has no markedly prognathous jaw and a flatter face than does *Australopithecus*. (T. Michael Keesey)

A. ramidus and *A. kadabba*. Table 10.2 gives details of these and other fossil hominin finds.

Australopithecus. Ardi-like species probably lived in the forests of Africa for hundreds of thousands or even a million years, extending their reach beyond the forests as the grasslands took hold. The next oldest fossils represent species that lived 4.2 to 1.1 million years ago, and they show dramatic changes from Ardi. Fossils of eight species of small hominins have been assigned to the genus *Australopithecus* (Table 10.2). These species differ from one another, and their exact phylogenetic relationships are not well understood. Some species are more ape-like, others have more human features such as a flatter face and arched bones in the feet. All australopithecines had small stature and small brains, although some evolved somewhat larger brains than Ardi, and more upright posture and walking. A remarkable trail of fossilized footprints is additional evidence of their hominin heritage. From these footprints it is clear that at least some *Australopithecus* walked fully upright. *Australopithecus* lived in a wide range of habitats including forests and open grasslands.

The move into open grassland may represent a significant evolutionary development, because once the need to move in trees is eliminated, selective pressure shifts toward more complete upright walking and running. A fossil specimen nicknamed Lucy represents the most well-known *Australopithecus* species, but Lucy is just one of many *Australopithecus* fossils (Figure 10.6). It seems clear that one *Australopithecus* species was the ancestor of the *Homo* genus, but at this point it is not possible to say if Lucy or some other *Australopithecus* species was our ancestor.

Paranthropus. You may still be thinking of human evolution as a straight line that begins with our common ancestors with chimps and ends with you. The fossil record shows that there were many side paths and dead-end branches in hominin evolution, just like there are with the evolution of any group of organisms. Some hominin species lived and thrived for hundreds of thousands of years or longer, but were not our direct ancestors. One group of distant hominin cousins is *Paranthropus*, meaning "beside humans." This group is characterized by large teeth, large, protruding jaws, and skulls that have enlarged sagittal crests. All of these specializations are related to a diet of tough plant fibers or nuts with hard shells. Large jaw muscles are needed to crush tough plants or crack nuts. Most researchers hypothesize that *Paranthropus* descended from *Australopithecus*. The group of species dates from about 3 to 1.5 million years ago. Although there are three species of *Paranthropus* (Table 10.2), the genus seems to be an evolutionary dead end that gave rise to no other group of hominins.

Homo. *Ardipithecus* and *Australopithecus* are clearly *not* human, and neither is *Paranthropus*, even though they have many human traits. Around 2.4 million years ago hominins started to emerge that look nearly like modern humans. Nine species of *Homo* are recognized, including *H. sapiens* (Table 10.2), although the exact number of species will certainly change as scientists find more fossils and analyze them more thoroughly. The emergence of the genus *Homo* was a major transition in human evolution. Any of the early *Homo* species could have survived to be the modern humans of today, or could have shared the modern world with us. But, today, for reasons not yet understood, *Homo sapiens* is the sole survivor of this history. All other so-nearly human species are extinct. Let's take a closer look at the evolution of various *Homo* species.

Table 10.2 Details of fossil hominids.

NAME	AGE	BRAIN VOLUME	LOCATION
Ardipithecus ramidus "Ardi"	5.8–4.4 MY	300–350 cm^3	Ethiopia
Ar. kadabba	5.8–5.2 MY	?	Ethiopia
Australopithecus anamensis	4.3–3.89 MY	?	Kenya
Au. afarensis "Lucy"	3.9–2.9 MY	400–500 cm^3	Ethiopia, Tanzania, Cameroon
Au. garhi	2.5 MY	?	Ethiopia
Au. africanus	3–2 MY	400 cm^3	South Africa
Paranthropus aethiopicus	2.8–2.3 MY	?	Ethiopia, Kenya
P. robustus	2–1 MY	515 cm^3	South Africa
P. boisei	2.3–1.4 MY	508 cm^3	Tanzania, Kenya, Ethiopia
Homo gautengensis	2 million–600,000 YA		South Africa
H. ergaster	1.9–1.49 MY		Kenya, South Africa
H. rudolfensis	1.9–1.88 MY		Kenya
H. habilis	1.83–1.5 MY	641 cm^3	Kenya, Tanzania
H. georgicus	1.8 MY	660 cm^3	Dminisi, Georgia
H. erectus	1.8 MY–300,000 or to 227,000 YA	variable: 931–1149 cm^3	Africa, Europe, Asia
H. floresiensis	1.1 MY–17,000 YA	380 cm3; possibly 417 cm^3	Indonesia
H. heidelbergensis	880,000–125,000 YA	?	England, France, Germany, Israel, Italy, Greece, Morocco, Spain, Hungary
H. antecessor	780,000 YA	?	Spain
H. rhodesiensis	640,000 YA	1267 cm^3	Ethiopia, Zambia, South Africa
H. sapiens (archaic)	500,000 YA		Europe
H. sapiens	270,000–present	1200 cm^3	Global distribution
H. neanderthalensis	175,000–32,000 YA	1420–1500 cm3	Europe, Middle East, Asia

The first *Homo* populations emerged around 2.4 million years ago and even the earliest species were more human than australopithecine. Later species were so like modern humans it is hard not to call them human. In this text we will identify all *Homo* species except *Homo neanderthalensis* and *Homo sapiens* by the term **Homo species**. These include *H. habilis, H. ergaster, H. erectus, H. heidelbergensis,* and others (Table 10.2). *Homo erectus* is the most commonly described fossil species, and you may see this name a lot in popular literature. It will be easier for you to see the early patterns of human evolution if we concentrate on trends in *Homo* evolution and avoid individual species names. Because *Homo neanderthalensis* and *Homo sapiens* are so recent and closely related, we will refer to these species by their full species names.

Even the earliest *Homo* species, that lived 2.5 million years ago, were tall—between 1.5 to 1.8 meters tall (5' to 6'). In fact, some *Homo* species were taller on average than modern *H. sapiens*. Regardless of height, early *Homo* species were very much like modern *H. sapiens* in body form and proportions. They were probably efficient runners, had sparsely haired bodies like ours, and were dark-skinned. The significant body differences from modern humans are in the skull. One trend in hominin evolution is reduction of tooth size. The earliest *Homo* species had larger teeth than the later species, and the teeth of *H. sapiens* are smaller than the teeth of other *Homo* species. These changes in tooth size are related to diet. In general, smaller teeth and smaller jaws reflect a softer diet—one that

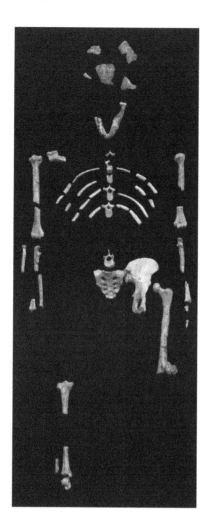

Figure 10.6: Skeleton of *Australopithecus afarensis*. Here is the ground-breaking 3.2-million-year-old partial skeleton called "Lucy." She is a small australopithecine.

requires less muscle power to dissect and mash food into portions that can be swallowed and digested.

Several other skull features changed systematically from early *Homo* species to *Homo sapiens*. Many *Homo* species have heavy brow ridges above the eyes that extend across the front of the skull (Figure 10.7). Brow ridges provide places where jaw muscles can attach; large brow ridges are associated with huge jaws and powerful molars needed to pulverize tough plant materials; smaller brow ridges, smaller jaws, and smaller molars are associated with softer foods that require less mechanical processing in the mouth before being swallowed. *Homo sapiens* has brow ridges, but they are much less pronounced. Over millions of years the *Homo* face has become flatter as the jaw becomes much smaller and recedes backward. Modern humans have a chin that supports the smaller jaw and teeth. Earlier hominin species have no chins or only slight chins.

One important trend in hominin evolution is an increase in brain size (Table 10.2). A modern human's brain has an average volume of about 1,400 cm³. In contrast, australopithecines had brains that were a bit more than a third of this size. Early *Homo* species had larger brains of about 700 cm³, while later *Homo* species had much larger brains—about 1,100 cubic centimeters. So, the conclusion seems to be that body size and proportions similar to humans emerged very early in the evolution of *Homo* species, while skull features and brain size moved more gradually toward traits that are typical for modern *H. sapiens*.

Presumably, a larger brain allows *H. sapiens* to achieve greater cognitive, linguistic, and intellectual abilities. There is no doubt that modern humans have greater intellectual skills than do other living great apes, although chimps and gorillas do make and use tools and can learn language. The abilities of *Homo sapiens* differ in degree from those of other apes, not in fundamental kind. The early *Homo* species did not achieve the cultural complexity of modern humans, but it is clear they shared many of our intellectual skills. For example, all *Homo* species made and used sophisticated stone tools. *Homo* species are not the only species to make and use tools, but *Homo* tool use has been widespread and common, and has evolved in complexity over the past 2.5 million years. The first stone tools emerge around 2.5 million years ago, and they are found associated with the earliest *Homo* species fossils (Figure 10.8). Australopithecines may have made stone tools, but *Homo* species certainly did. The earliest tools were simple, but they were effective and for the last 2 million years they are found associated with *Homo* species in Africa and across Asia and Europe. These tools were made by holding a rock of flint or other suitable stone in one hand, and striking the flint *just right* with a stone held in the other hand. The strike threw off a flake, and left a sharp edge in the flint. The flint stone usually had a rounded base that made holding it easy, and the sharp edge could be used to cut, scrape, chop, and do all sorts of useful things. One of the striking patterns in *Homo* evolution is the increased sophistication of stone tools over time. Even while the simple flaked core tool was used over a million years or more, by species across the Old World, by 1.8 million years ago more sophisticated tools appeared in Africa and later in other regions. These later tools were made by striking flakes off more than one side of the core flint stone. This produced a true knife or axe, with sharp edges and a sharp tip. Stone handaxes made using this technique were a significant advance.

(a)

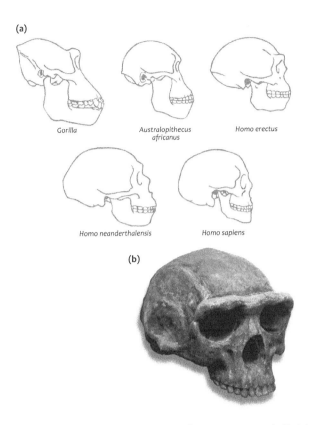

Gorilla

Australopithecus
africanus

Homo erectus

Homo neanderthalensis

Homo sapiens

(b)

Figure 10.7: Primate skulls in profile and reconstruction of *Homo erectus* skull. (a) Note the heavy brow ridges in the skulls of gorilla, *Homo erectus*, and *Homo neanderthalensis*. (b) This reconstructed *Homo erectus* skull shows the heavy brow ridges typical of the species. (a: V.P. VOLKOV; b: THOMAS ROCHE)

From there even more sophisticated tools evolved. For example, early *Homo sapiens* made much finer and sharper tools, and learned to use the discarded flakes as very delicate tools. Thus, like the biological evolution of brain and skull features, the cultural evolution of tool use changed gradually over the 2.5 million years of *Homo* history. It is tempting to conclude that the increase in brain size in *Homo* species is responsible for accumulation of human cultural and cognitive sophistication.

One distinctive feature of *Homo* species is their tendency to travel. In their long years on Earth some *Homo* species, especially *Homo erectus* migrated to many parts of the world. Fossils similar to *Homo erectus* have been found in China, Eurasia, Europe, and even in Indonesia. Early *Homo* species began migrating out of Africa at least 1.5 million years ago and persisted in some locations as recently as 50,000 years ago and perhaps even more recently. Contemporary humans take pride in our long history, but our history is dwarfed by that of *Homo erectus* and its cousins who were successful hominins for nearly 2 million years.

Of course one *Homo* species stands out in the interest of many people: *Homo neanderthalensis.* Neanderthals are our closest known relatives, so it is worth spending time to get to know them. Neanderthals evolved from an early *Homo* species, perhaps *H. heidelbergensis,* that migrated to Europe

Figure 10.8: 1.2 million-year-old handaxe. This handaxe from the Olduvai Gorge in Tanzania will fit into the palm of your hand. Its sharp edge could slice flesh or chop wood. (BABELSTONE)

Figure 10.9: Neanderthal skull. This replica of a Neanderthal skull found in St. Michael's Cave on Gibraltar clearly shows the swelling at the rear of the Neanderthal skull that contributes to the increased brain volume of Neanderthals. (NATHAN HARIG)

about 500,000 years ago. Neanderthals appeared in Europe about 120,000 years ago, and persisted in Europe until 25,000 to 30,000 years ago. What were they like?

One question is what did Neanderthals look like? In Europe the Neanderthal fossil record is rich and much is known about their physical traits. Most Neanderthal fossils have distinct features from *Homo sapiens*, but there is a lot of overlap. Neanderthal skulls have sloped foreheads and a bulge in the rear of the skull (Figure 10.9). Their noses were probably broader, and their bodies were squatter than those of modern humans. On average Neanderthal men were about 65 cm tall (about 5.5 ft), while the women were about 60 cm tall (5 ft). Stronger than a typical *Homo sapiens* of the same size, Neanderthals had relatively longer arms, and larger chests. You might think that Neanderthals had brains somewhat smaller than *Homo sapiens,* but they did not. Modern human brains average 1,200 to 1,400 cm^3, but Neanderthal brains averaged about 1,500 cm^3. This picture of Neanderthals paints them as physically very similar to modern humans, but recognizably different. But, what were Neanderthals like culturally? Did they live like us, or at least like our *Homo sapiens* ancestors? How did they become extinct?

Popular culture depicts Neanderthals as brutish cavemen with little intelligence. The fossil record suggests this picture is far from accurate. From modern-day Israel, to Siberia, and west to Great Britain, Neanderthals and their immediate ancestors had a widespread and successful culture for nearly half a million years. Sophisticated stone tools were made and used by Neanderthals. While these tools were not as varied as those used by early *Homo sapiens*, they were effective in cutting, butchering, and hunting. Body decorations, burials, use of fire, and complex living structures also indicate a sophisticated culture. Certainly the *Homo sapiens* who arrived in Europe about 50,000 years ago had a more varied culture, more rich in art and use of symbols, but Neanderthals also had a rich culture and a strong social infrastructure, the details of which we are still learning. ...

One approach to finding out if Neanderthals had speech is to look at their anatomical structures. Some fossilized anatomical features can be used to infer the size of tracts of nerves that supply the tongue and the structure and size of their vocal tracts. Neanderthals had anatomical structures similar, but not quite the same as those of humans. Some researchers have looked at Neanderthal anatomy and concluded that they could not make the sounds that we can. Many others have concluded that while Neanderthal anatomy is not the same as *Homo sapiens*, they could produce many if not all the sounds that we do. The issue is not yet resolved. There is another interesting approach, however. Recent DNA studies have shown that humans have at least one very distinct gene, called *FoxP2* that is involved in and critical for language. All mammals have this gene, but the human gene is different in three critical sites from the gene of other mammals. Here is an astounding finding: researchers have been able to isolate and sequence Neanderthal DNA that includes this gene. You should be aware that there are many challenges to studying Neanderthal DNA and so this result *could* be contradicted in the future. Still, these researchers found that Neanderthals carried the modern human version of the *FoxP2* gene, not the chimp or common mammalian version. So, it is possible that Neanderthals could use language and talk. Whether they did or not is another question.

In the complex and varied history of *Homo* species only one remains, *Homo sapiens*. All others, including Neanderthals, are extinct. But Neanderthals co-existed in Europe with *Homo sapiens* for at least 10,000 to 20,000 years. Did the two species interbreed? In Europe some fossil remains look "intermediate" between *Homo sapiens* and Neanderthals, but it is hard to interpret these results because there is natural variation in both populations. Some researchers have turned to DNA to answer this question. DNA from Neanderthals has been isolated, analyzed, and compared with modern human DNA. ...

EDITOR'S NOTE

Studies published in the last couple years indicate that Neanderthals were, in fact, a separate species from *Homo sapiens*, and molecular data suggests that non-African humans today carry with them somewhere between 1 and 4 percent Neanderthal DNA, indicating that the two species interbred while co-mingling in Europe and Asia.

WHERE IS THE "MISSING LINK" IN HUMAN EVOLUTION?[5]

Ever since Darwin rightly concluded that humans evolved from ancient apes—but not living apes of course—the popular press has asked for some evidence of a "missing link" that would connect apes to humans. Most people expect to find one or two clear intermediate forms between modern chimps and modern humans—forms that represent the transition from ape to human. The science of human evolution has at least two answers to this quest. First, intermediate forms between the chimp/human ancestor and modern chimps and humans are abundant, but diverse. Second, the idea of a straight line of evolution from the common chimp/human ancestor to modern humans is not supported by the fossil record. Many different kinds of "humans" have lived over the past five million years, each with some traits that are intermediate between chimps and humans, some that are mostly human or chimp, and some traits that are different from human or chimp. The lesson here is that popular conceptualizations of how the world works are often naïve and misleading. The whole idea of a "missing link" misses the point of how evolution happens, and in particular the history of human evolution. The idea of a "missing link" assumes a straight line of evolution from an ancestor to a modern form. By its nature evolution produces many different species, some of which survive, some of which do not, and some of which evolve into other species. If you look back in time to find the ancestors of any living species you will find many candidates, each of which has some "missing link" traits. Scientists build phylogenetic trees that describe this evolutionary history, and the intermediate forms between different groups. This is true in human evolution as well. Scientists have an increasingly clear picture of what kinds of hominins lived at different times, and how they compare with modern humans. All of these ancient hominins have transitional traits. As more discoveries are made, we should be able to identify transition species between the ancestors we share with chimps and ourselves.

DO TOOLS MAKE US HUMAN?

For as long as people have talked and written about themselves, they have pondered what makes people different from other animals. Some people have answers rooted in religious beliefs; others have cultural or social answers.

---

5   Jenner, Jan; Presson, Joelle, "The Evolution of *Homo sapiens*," *Biology: The Tapestry of Life*, pp. 443-444. Copyright © 2014 by Cognella, Inc.. Reprinted with permission.

Scientists propose testable hypotheses about what makes us human, and look for objective, measurable traits with which to test those hypotheses. One idea that often comes up is that humans are "tool users" and other animals are not. It seems obvious that humans imagine, design, manufacture and use amazing tools. *Homo* species have been doing this for two million years, and no other ancient genus or species have fossils so clearly associated with stone tools. Does all this mean that humans are the only tool makers? The answer turns out to be no. In the 1970s Jane Goodall made history by recording the use of tools by chimpanzees. Chimps use all kinds of tools: stones to smash nuts, twigs to stab tasty insects, sticks to dig up roots. Chimps even fashion some of these tools: they modify stems, sticks, vines and stones to do a specific job. Since this discovery other animals have been found to use tools. Crows use complex, sculpted twigs to reach inaccessible food. Dolphins use sponges to help uncover food sources. Amazingly, an inventive octopus has been caught on film stacking up coconut shells from the sea floor and carrying them away to use later as shelters. Whales make thick curtains of bubbles to confine the fishes or shrimp they eat. The list of animals known to use tools seems to get larger every year. These findings lead scientists to question whether using tools is really a defining feature of humans. What do *you* think? Is human tool use fundamentally different from the use of tools by other animals?

GLOSSARY

CHAPTER 1 GLOSSARY

Anthropology - The study of human cultures and societies.

Archaeology - The study of ancient humans and the artifacts they left behind.

Biology - The study of living things.

Catastrophism - The long-standing belief that the Earth has been shaped by catastrophic events, such as giant floods and earthquakes, which were caused by a divine being. This is roughly the opposite of uniformitarianism (see below).

Fossil record - The fossil record consists of all the fossils we have found so far, and is often described as a chronological sequence of fossil assemblages as they appear throughout the past.

Geology - The study of the Earth, its history and the processes that shape it.

Meteorology - The study of weather and the atmosphere.

Oceanography - The study of the oceans.

Paleontology - The study of the fossil record.

Science - The human activity of seeking a natural explanation for things that we observe.

Uniformitarianism - The belief that the Earth is shaped by slow, ongoing processes. A tenet of this notion is that *the present is the key to the past*. This means that the phenomena we witness today, such as erosion, sedimentation, and tectonic movement, occur at rates that have been more or less constant through time. This theory contradicts catastrophism (see above).

CHAPTER 2 GLOSSARY

Abrasion - A process where rock is ground down and chipped away by the collisions of, or sliding against, other rocks.

Angular - Having sharp edges and corners, such as a freshly broken rock.

Anoxia - Lacking sufficient oxygen, such as the ponds and puddles in a swamp.

Anthropocene - A new epoch proposed that marks the period of time during which humans have permanently altered the planet.

Body Fossil - A fossil of an organism's actual body, such as a skeleton or carbon imprint.

Breccia - A poorly sorted clastic sedimentary rock that contains angular grains.

Capacity - A measurement of how much sediment a river is capable of moving.

Carbonization - A process where the volatile components of fossil tissue diffuse away, leaving behind a concentrated film of carbon.

Cementation - A type of lithification in which a mineral, typically deposited by groundwater, forms between sediment grains, gluing them together to form a solid rock.

Chert - A silica-based chemical/biochemical sedimentary rock that typically forms in the deep ocean.

Clastic - Containing or composed of fragments of pre-existing rocks (sediments).

Coal - A biochemical sedimentary rock formed from the compaction and decomposition of ancient organic material, typically plants.

Cliff Retreat - A phenomenon where a cliff is eroded over time, causing the edge of the cliff to migrate backward.

Compaction - A type of lithification in which sediments such as clay adhere to one another to form a more solid rock.

Competence - A measure of the maximum size of a rock a river is capable of moving.

Conglomerate - A poorly sorted clastic sedimentary rock that contains rounded grains.

Coprolite - Fossilized feces.

Correlation - The practice of identifying rocks of similar age and/or origin, using a variety of techniques.

Desert Pavement - A concentration of pebbles and cobbles on the desert floor that results when the smaller grains, such as clay, silt, and sand, are blown away by the wind.

Desiccation - The process of drying out.

Dissolution - A chemical reaction in which solid materials (called *solute*) dissolve into a *solvent* such as water or acid. This is a type of chemical weathering that affects limestone most commonly.

Erosion - When weathered sediments are carried away by gravity or some sort of current, such as water or wind.

Erratics - Large rocks carried and deposited by glaciers.

Evaporite - A type of chemical sedimentary rock formed when dissolved minerals precipitate during the evaporation of water. Salt (halite) and gypsum are two examples of evaporites.

Exfoliation - The peeling or sloughing off of layers.

Frost Wedging - A type of physical weathering in which water in the cracks of rocks freezes and expands, breaking the rock apart.

Glacial Polish - A smooth surface created on bedrock that has been ground down by moving glaciers.

Glacier - Glaciers form when large, deep piles of snow compact into solid ice that is capable of flowing outwards and downwards under the influence of gravity.

Hydrolysis - A chemical reaction in which water breaks down minerals. For instance, the hydrolysis of feldspar, in the presence of acidic rain, results in the formation of clay minerals.

Inclusion - Something embedded in something else. For example, a pebble in a conglomerate is an inclusion. In fact, any grain of sediment in any clastic sedimentary rock is considered to be an inclusion.

Joints - Planar, sometimes evenly-spaced, fractures in a rocky outcrop.

Karst Topography - A landscape controlled by the dissolution of limestone bedrock, often marked by sinkholes and steep-sided spires of rock.

Limestone - A common chemical/biochemical sedimentary rock made primarily of the mineral calcite.

Lithification - A process where new sedimentary rocks are formed. Two common types of lithification are *cementation* and *compaction*.

Oxidation - A chemical reaction in which water removes electrons of a substance. A common type is the oxidation of iron, resulting in reddish minerals such as *hematite*.

Permineralization - A type of fossilization where minerals grow within the pores and cavities of the original organic tissue.

Playa - A dried-up lake bed.

Relative Dating - Any method that helps us compare the ages of rocks. For instance, the *principle of superposition* states that a stack of sedimentary layers will be youngest at the top and oldest at the bottom.

Replacement - When the original tissue of a fossil is completely replaced by new minerals.

Rounding - A measurement of how much a rock has been smoothed by weathering.

Sandstone - A common clastic sedimentary rock composed primarily of sand.

Sediments - Loose rocks that result from the weathering of solid bedrock.

Shale - A fine-grained clastic sedimentary rock composed mostly of clay.

Sorting - A measurement of how much size variation there is among the sediments in a rock.

Striations - Scratch marks created by the abrasion of rocks. For instance, glaciers can leave behind striations because rocks along the bottom of the ice get dragged across the landscape.

System - Rocks from the past arranged in chronological order.

Thermal Contraction and Expansion - The decrease and increase in the volume of a material due to temperature fluctuations.

Till - Sediments deposited by glaciers.

Trace Fossil - A fossil left behind by an organism that doesn't include the actual organism's body, such as a footprint.

Time-Equivalent - Rocks or other geologic features are found to be of similar age.

Travertine - A form of limestone typically produced along springs.

Unconformity - As seen in stacks of sedimentary rocks, these are breaks in the geological record where there are no rocks to represent particular spans of time from the past.

Unloading - The removal of overlying rocks due to erosion or tectonic movement.

Ventifact - A rock faceted and polished by the abrasion of wind-blown sand.

Weathering - Any process that breaks down rocks into smaller particles.

Organic - When living things weather rocks, such as plant roots.

Physical - Also called *mechanical*, when rocks are physically weathered, typically through abrasion.

Chemical - A type of weathering where rocks are broken down chemically, such as through dissolution, oxidation, and hydrolysis.

CHAPTER 3 GLOSSARY

Arc - A long, arc-shaped line of volcanoes that form along subduction zone plate boundaries.

Asthenosphere - A zone of soft, partially-melted rocks within the upper mantle that underlies the lithosphere.

Collisional Plate Boundary - A type of convergent boundary where two continental land masses collide, forming a tall, wide mountain range (ex. Himalayas).

Conduction - The transfer of heat through a solid medium.

Convection - The transfer of heat through a liquid or gas, aided by the flow of currents.

Convergent Plate Boundary - A plate boundary where two lithospheric plates are crashing into each other.

Continental Crust - Crust that generally forms land masses and is thicker and more buoyant than oceanic crust.

Continental Drift Hypothesis - A hypothesis stating that the continents have been moving around relative to each other and had once formed a supercontinent, Pangaea.

Divergent Plate Boundary - A plate boundary where oceanic crust is formed and the rocks on either side are being pulled apart.

Differentiation - The separation or sorting of materials into distinct zones.

Felsic - Containing relatively high amounts of silica and very little to no iron or magnesium (opposite of *mafic*).

Ferromagnesian - Also known as "mafic," this refers to minerals and rocks that are rich in iron and magnesium, and are typically dense, dark, and have high melting temperatures.

Fold and Thrust Belt - A belt of folded and faulted rocks that forms along the leading edge of the overriding plate of a subduction zone.

Geothermal Gradient - The increase in the temperature of progressively deeper crustal rocks.

Island Arc - A chain of volcanic islands that form along subduction zones.

Lithosphere - The rigid outer layer of the Earth, composed of the crust and the solid uppermost mantle, that sits directly atop the asthenosphere.

Mafic - (See *Ferromagnesian.*)

Mantle Wedge - The portion of mantle found between the overriding and downgoing plates of a subduction zone.

Pangaea - An ancient supercontinent that had broken apart to form today's current land masses.

Plate Tectonics - A theory that states the Earth's lithosphere is broken into multiple plates that drift around and interact with each other along plate boundaries.

Rift Valley - A valley that forms when a continental landmass first begins to diverge.

Seafloor Spreading - This occurs when the ocean floor has a divergent boundary and grows as a result.

Subduction Zone - A type of convergent plate boundary where one plate gets shoved beneath another plate.

Tectonic Plate - A fragment of the lithosphere that moves around independent of the other fragments.

Transform Plate Boundary - A type of plate boundary where the rocks on either side slide past each other. This is a result of shear forces and is best exemplified by the San Andreas Fault.

CHAPTER 4 GLOSSARY

Adaptive Radiation - When organisms diversify and end up filling new ecological niches.

Allopatric Speciation - When subsets of a population of a species become isolated from the rest and develop their own unique characteristics.

Artificial Selection - The human activity of forcing evolution on a species by selecting particular traits to be accentuated through inbreeding.

Biostratigraphic Correlation - A method where rocks are found to be of similar age based on the fossils they contain.

Biozone - A set of rock strata defined by the fossils it contains.

Concurrent Range Zone - A set of rock strata defined by the overlapping of two or more individual range zones.

Faunal Succession - The pattern we see of fossil assemblages succeeding one another through time in a regular and determinable order.

Guide Fossil - A fossil used to establish the estimated age of the rock it is found in.

Homologous Structure - A trait or set of traits that appear in a group of organisms inferred to have a common ancestor. A good example would be the arrangement of bones found in vertebrates.

Lithostratigraphic Correlation - Identifying rocks of similar age based on the characteristics of the rocks themselves.

Microevolution - Small-scale evolutionary changes.

Natural Selection - The phenomenon where certain members of a species have traits that enable them to reproduce more successfully than others. As a result, these beneficial traits increase in abundance whereas negative traits are weeded out of the population. This is the mechanism that allows evolution to take place.

Phyletic Gradualism - The observation that some species appear to evolve slowly and progressively through time.

Phylogenetic Tree - A diagram, typically illustrated as a set of branching lines, that illustrates the inferred evolutionary relationships between different species.

Punctuated Equilibrium - The observation that some species evolve quickly, then reach a sort of equilibrium with their environment and undergo a subsequent period of stasis.

Range Zone - The span of geological time during which a species is present.

Species - A group of organisms with many common characteristics that is capable of interbreeding.

Stasis - Without change.

Vestigial - Some sort of trait or behavior believed to be an evolutionary remnant of an organism's ancestors (ex. human appendix).

CHAPTER 5 GLOSSARY

Allele - One of the variations of a particular trait. If the trait is "eye color," then an allele could be, for example, "blue eyes."

Archaea - A group related to bacteria but slightly more similar to eukaryotes than bacteria. They tend to be capable of surviving in extreme environments other forms of life could not inhabit.

Bacteria - A broad group (domain) of simple, single-celled prokaryotes.

Clade - A cluster of lineages that branch off from a common ancestor, often shown as a "tree" on a graph.

Convergent Evolution - When distantly related organisms develop similar traits, such as the wings seen in insects, birds, bats, and flying reptiles.

Directional Selection - A type of natural selection where a trait, such as size, changes over time.

Divergent Evolution - When a species evolves progressively more and more different from its ancestors.

DNA - Deoxyribonucleic acid, a long chain of nucleotides that occurs in specific patterns that determine how an organism develops, grows, and reproduces.

Eukaryote - A more complex type of cell than prokaryotes. Eukaryotic organisms have membrane-bound organelles, reproduce sexually, and include many multicellular forms of life.

Gene - A sequence of nucleotides in a strand of DNA that determines particular traits.

Genotype - Refers to the genes that determine particular traits.

Linnaean Classification System - A system of organizing living things into a nested, hierarchical set of groups that range from very broad in scope (ex. kingdoms) to very specific (ex. species).

Mass Extinction - A devastating event where a significant portion (~75%, give or take) of the world's species is wiped out.

Monophyletic - A group of organisms that consists of a common ancestor, and ALL of its descendants.

Parallel Evolution (or **parallelism**) - When closely related organisms develop similar traits, such as those between placental and marsupial mammals.

Parapatric Speciation - When two or more species develop when the region in which they live exhibits different selective forces depending on geographical location.

Paraphyletic - A group of organisms that consists of a common ancestor and only SOME of its descendants.

Peripatric Speciation - When a new species forms when a subgroup of an ancestral species, typically on the periphery of the population, becomes isolated and is forced to adapt to a new environment.

Phenotype - Refers to the traits of an individual that can be described through observation.

Phenotypic Classification - The practice of classifying organisms into groups that share similar characteristics and traits.

Phylogenetic Classification - The practice of classifying organisms into groups that are related genetically and share common ancestors.

Polyphyletic - A group of organisms that share particular traits but do not share a direct common ancestor.

Prokaryote - A simple type of cell that does not contain membrane-bound organelles and reproduces asexually. Archaea and bacteria are both prokaryotes.

Shared Ancestral Traits - Traits shared by organisms that are derived from a common ancestor.

Shared Derived Traits - Recently-developed traits that are exclusively developed within a particular population.

Speciation - When a new species branches off from an ancestral species.

Stabilizing Selection - A type of natural selection that weeds out organisms that exhibit extreme versions of a particular trait.

Sympatric Speciation - When two or more species form within the same geographical area, driven apart by extremes within the environment.

CHAPTER 6 GLOSSARY

Abiogenesis - The formation of life from nonliving matter.

Accretion - A process in which materials clump together.

Anaerobic - Lacking or not requiring free oxygen.

Banded Iron Formations (BIFs) - Iron-rich sedimentary layers that formed in the oceans when dissolved iron suddenly oxidized.

Chloroplast - An organelle in plants and protists that collects sunlight and converts it to food and energy through photosynthesis.

Circumstellar Disc - A disc-shaped cloud of dust and gas that once surrounded the sun and later accreted into the planets.

Condensation - A process in which a gas changes phase to become a liquid.

Cyanobacteria - Photosynthesizing prokaryotes, one of the earliest-recognized forms of life found in the fossil record.

Ediacaran Fauna - A group of multicellular, soft-bodied fossils from the Late Precambrian that represent the earliest known macroorganisms.

Endosymbiosis - A symbiotic relationship where one organism lives inside of another.

Golgi Apparatus - An organelle found in most eukaryotic cells that helps to organize and synthesize complex molecules such as proteins and lipids.

Great Oxygenation Event - A sudden increase in the amount of free oxygen during the Proterozoic, likely due to the increase in photosynthesis during this time.

Macrofossil - A fossil large enough to be seen with the naked eye.

Mitochondria - An organelle that converts nutrients into energy for the cell.

Organelles - A specialized structure within a cell, such as *mitochondria*.

Period of Heavy Bombardment - A period of intense meteoric activity that occurred nearly four billion years ago.

Protobionts - A membrane-bound cluster of organic molecules that may be a precursor to the prokaryotes.

Snowball Earth Theory - The theory that during the Late Proterozoic, the Earth underwent several extreme global ice ages.

Spontaneous Generation - An antiquated theory that living things are derived from matter in which they reside. For instance, it was once thought that flies formed directly from putrid meat.

Stromatolite - A layered structure consisting of sediments and cyanobacteria that grow along shorelines, and represent some of the earliest fossils.

CHAPTER 7 GLOSSARY

Amniote - Some tetrapods, such as reptiles, birds, and mammals, which lay their eggs on land or carry them within the mother.

Arborescent - Having tree-like characteristics.

Benthic - Pertaining to the sea floor.

Cambrian Explosion - A sudden increase in the diversity and complexity of life that occurred during the Cambrian Period.

Chordata - A phylum of organisms composed primarily of vertebrates; they have bilateral symmetry and a central nerve cord.

Epifauna / Epiflora - Organisms that live on the surface of the ocean floor.

Fishapod - An informal name given to transitional creatures that bridge the evolutionary gap between fish and tetrapods (four-limbed vertebrates).

Infauna - Organisms that live within the sediments of the ocean floor, such as those that burrow.

Lophophore - A horseshoe- or ring-shaped tentacle used by some aquatic invertebrates to collect nutrients from the water.

Mobile - Describing benthic organisms that are free to crawl or slither across the ocean floor.

Nektonic - Having the ability to swim or otherwise move on one's own through the water.

Pelagic - Living up in the water column, above the ocean floor but below the surface of the ocean.

Photic Zone - The upper layer of the ocean where light penetrates and photosynthesis can occur.

Protorothyrid - Known as a "stem reptile," this is a Permian reptile from which mammals, dinosaurs, and birds evolved.

Suture - An irregular line seen on the surface of ammonite shells that indicates the intersections of the inner septa with the outer wall of the shell.

Synapsid - A group of mammals and their direct ancestors that share a similar structure of the skull.

Tetrapod - A four-limbed vertebrate.

CHAPTER 8 GLOSSARY

Albedo - A measurement of the percent of light that is reflected off a surface.

Epeiric Sea - A shallow sea that forms atop the continental shelves during marine transgressions.

Extremophiles - Organisms capable of surviving in extreme conditions that were once thought to be unable to harbor life.

Gondwana - A supercontinent in the Southern Hemisphere that once included today's southern continents as well as India.

Laurasia - A supercontinent in the Northern Hemisphere that once included today's northern continents.

Panthalassa - A large ocean that occupied the other side of Earth from Pangaea.

Orogenic - Relating to mountains or mountain-building processes.

Regression - A lowering of sea level that tends to expose the continental shelves.

Rodinia - A supercontinent that predates Pangaea; it broke apart around 750-600 million years ago.

Tethys Sea - A sea that once existed along the eastern margin of Pangaea.

Transgression - A rise in sea level wherein the coastal margins of the continents are inundated.

Ultramafic - A rock composition typical of the mantle where the majority of the rock is composed of mafic minerals.

Western Interior Seaway - A body of water that stretched across most of central North America during the Late Cretaceous and also for a time in the Early Cenozoic.

CHAPTER 9 GLOSSARY

Artiodactyl - A hoofed animal with an even number of toes (two or four) on each foot.

Cetaceans - Whales, dolphins, and porpoises.

Living Fossil - An organism living today that closely resembles its ancestors from the fossil record.

Marsupial - A mammal group that gives birth to live but underdeveloped young that further develop in the mother's pouch.

Milankovitch Cycle - A cycle of global cooling and warming that results from fluctuations in the Earth's axial tilt, direction of axis, and distance from the sun.

Monotreme - An egg-laying mammal (e.g., platypus).

Multituberculates - An extinct group of rodent-like mammals.

Paleocene/Eocene Thermal Maximum (PETM) - A sudden spike in global temperatures that occurred at the boundary of the Paleocene and Eocene epochs.

Perissodactyl - A hoofed animal with an odd number of toes on each foot (one, three, or five).

Physiognomy - The study of the different shapes something can take, such as the shapes of leaves, in this context.

Placental - A type of mammal that gives birth to well-developed live young after incubating them within the mother's abdomen.

Serrate - Having a rough or "toothed" edge.

Tuff - A volcanic rock that forms when ash is deposited in layers on the ground surface and later lithifies.

Ungulate - A hoofed animal.

CHAPTER 10 GLOSSARY

Hominid - All apes and humans belong to this group.

K/T Event - The mass extinction event that occurred at the end of the Cretaceous (K) and ushered in the Tertiary Period (T). Note, we use Paleogene now instead of Tertiary, and this event is also referred to as the K-Pg event.

New World - Pertaining to South and Central America, in the context of primates.

Old World - Pertaining to Africa and Asia, in the context of primates.

Primate - An order that includes all the monkeys, apes, and humans.

IMAGE CREDITS

CHAPTER 1

CHAPTER 2

CHAPTER 3

CHAPTER 4

CHAPTER 5

CHAPTER 6

CHAPTER 7

CHAPTER 8

CHAPTER 9

CHAPTER 10

GEOLOGIC TIME SCALE

EON	ERA	PERIOD		EPOCH	MILLIONS OF YEARS AGO
PHANEROZOIC	CENOZOIC	QUATERNARY		HOLOCENE	
				PLEISTOCENE	.011
		TERTIARY	NEOGENE	PLIOCENE	2.5
				MIOCENE	5.3
			PALEOGENE	OLIGOCENE	23.0
				EOCENE	34.0
				PALEOCENE	55.8
	MESOZOIC	CRETACEOUS			65.5
		JURASSIC			145.5
		TRIASSIC			199.6
	PALEOZOIC	PERMIAN			251
		CARBONIFEROUS	PENNSYLVANIAN		299
			MISSISSIPPIAN		318
		DEVONIAN			359.2
		SILURIAN			416
		ORDOVICIAN			443.7
		CAMBRIAN			488.3
PRECAMBRIAN	PROTEROZOIC	NEOPROTEROZOIC			542
		MESOPROTEROZOIC			1000
		PALEOPROTEROZOIC			1600
	ARCHEAN	LATE			2500
		EARLY			3200
	HADEAN				4000
					4600

INDEX

CPSIA information can be obtained
at www.ICGtesting.com
Printed in the USA
FSHW021752271219
65492FS